Discrete Mathematics Using a Computer

Springer
London
Berlin
Heidelberg
New York
Barcelona
Hong Kong
Milan
Paris
Singapore
Tokyo

Cordelia Hall and John O'Donnell

Discrete Mathematics Using a Computer

 Springer

Cordelia Hall, PhD
John O'Donnell, BS, MS, PhD

Department of Computing Science, University of Glasgow,
17 Lilybank Gardens, Glasgow G12 8QQ, UK

ISBN 1-85233-089-9 Springer-Verlag London Berlin Heidelberg

British Library Cataloguing in Publication Data
Hall, Cordelia
    Discrete mathematics using a computer
    1. Computer science – Mathematics
    I. Title   II. O'Donnell, John T.
    004'.0151
ISBN 1852330899

Library of Congress Cataloging-in-Publication Data
Hall, Cordelia, 1955-
    Discrete mathematics using a computer / Cordelia Hall and John
    O'Donnell.
        p.   cm.
    Includes bibliographical references and index.
    ISBN 1-85233-089-9 (alk. paper)
    1. Mathematics—Data processing.   I. O'Donnell, John, 1952-   .
    II. Title.
    QA76.95.H35   2000   99-26380
    510'.285—dc21   CIP

© Springer-Verlag London Limited 2000
Printed in Great Britain

Typesetting: Camera ready by authors
Printed and bound at the Athenæum Press Ltd., Gateshead, Tyne & Wear
34/3830-543210 Printed on acid-free paper SPIN 10696861

This book is dedicated to our parents:
David Hall, Patricia Hall, Estelle O'Donnell,
and to the memory of John A. O'Donnell.

Preface

Several areas of mathematics find application throughout computer science, and all students of computer science need a practical working understanding of them. These core subjects are centred on logic, sets, recursion, induction, relations and functions. The material is often called *discrete mathematics*, to distinguish it from the traditional topics of *continuous mathematics* such as integration and differential equations.

The central theme of this book is the connection between computing and discrete mathematics. This connection is useful in both directions:

- Mathematics is used in many branches of computer science, in applications including program specification, data structures, design and analysis of algorithms, database systems, hardware design, reasoning about the correctness of implementations, and much more;

- Computers can help to make the mathematics easier to learn and use, by making mathematical terms executable, making abstract concepts more concrete, and through the use of software tools such as proof checkers.

These connections are emphasised throughout the book. Software tools (see Appendix A) enable the computer to serve as a calculator, but instead of just doing arithmetic and trigonometric functions, it will be used to calculate with sets, relations, functions, predicates and inferences. There are also special software tools, for example a proof checker for logical proofs using natural deduction.

The software for the book uses Haskell, a mathematically-oriented programming language which is particularly well suited for discrete mathematics. It is assumed that the reader has no prior knowledge of Haskell; everything needed is covered in Chapter 1. A further discussion of the role of computing in the textbook, and the reasons for the choice of Haskell, is given below.

Chapter 1 introduces all the basic notations of Haskell that are used in the book. It is not necessary to read all of Chapter 1 before embarking on the rest of the book; it would be reasonable to start with Chapter 2, and to refer back to Chapter 1 as needed. However, it is useful to begin with a quick skim through Chapter 1 in order to get an overall impression of what is covered there. Only basic expressions and functions are needed in this book, and more advanced parts of Haskell are not covered here. The Further Reading section of Chapter 1 refers you to books that cover the full language.

Chapter 2 covers propositional logic, which is one of the foundations for formal reasoning in computer science. First the language of propositional logic is introduced, including a discussion about translating between English and formal propositions. Then three major systems for logical reasoning are introduced: truth tables, natural deduction with logical inference rules, and Boolean algebra.

Chapter 3 adds predicates and quantifiers to propositional logic, producing predicate logic. This system is frequently used to state properties of computer programs and to prove correctness of algorithms. Again, three methods of reasoning about predicates are presented: expanding formulas into equivalent propositions, logical inference and equational laws.

The basic concepts of set theory are covered in Chapter 4. The operations on sets are defined, and then demonstrated with simple software tools. The most important theorems about set operations are stated as equational laws, and applications to computing are used as examples.

Many structures in computer science are defined recursively. Chapter 5 introduces recursion as a technique for writing function definitions, and it is applied to lists and trees. Chapter 6 then develops a closely related technique, inductive definition of sets. Induction is the mathematical technique for proving properties of functions and sets that are defined recursively, and this is the subject of Chapter 7.

Chapter 8 defines relations, which are essential for many applications. It explains the basic properties of relations, and defines order relations and equivalence relations. Functions are a special kind of relation, and Chapter 9 covers this topic.

Many examples of the application of discrete mathematics to computing are given throughout the book, but these examples must necessarily be kept small. To give a realistic picture of how mathematics is used in large scale practical applications, Chapter 10 is devoted to one in-depth case study: the design of combinational digital circuits. Basic logic gates and circuits are described, and Boolean algebra is used to describe their behaviour. A larger scale problem—the specification and correctness proof of a general n-bit binary addition circuit—is then introduced and solved. This case study brings together many of the topics of the book, including Boolean algebra, functions, recursion, higher order functions and induction, in order to solve an interesting and practical computing problem.

Software Tools for Discrete Mathematics

A central part of this book is the use of the computer to help learn the discrete mathematics. The software (which is free; see below) provides many facilities that aid the student in learning the material:

- Logic and set theory have many operators that are used to build mathematical expressions. The software allows the user to type in such expressions interactively and experiment with them.

- Predicate logic expressions with quantifiers can be expanded into propositional logic expressions, as long as the universe is finite and reasonably small. This makes the meaning of the quantifiers more concrete and helps the development of intuition.

- Students frequently misuse expressions in logic and set theory; a typical error that arises frequently is to write an expression that treats $A \subseteq B$ as a set rather than a Boolean value. The software tools will immediately flag such mistakes as type errors. Teaching experience shows that many students will have long-lasting misconceptions about basic notations without immediate feedback.

- A formal proof checker for natural deduction is provided. This allows students to find errors in their proofs before handing in exercises, and it also provides a quick and effective way for the instructor to check the validity of large numbers of proofs. Furthermore, the automated proof checker underscores the nature of formal proof; vague or ill-formed proofs are not acceptable.

- Using a proof checker gives a deeper appreciation of the relationship between discrete mathematics and computer science. The experience of debugging a proof is much like debugging a computer program; the proof checker is itself a computer program (which the students can read if they wish to); proof checking software makes formal proof feasible for larger scale problems.

- The techniques of recursion and induction are applied directly and formally to function definitions which the student can execute.

The programming language used in the book is Haskell 98 (which will be called simply Haskell). This is a standard pure functional language with excellent support. Several implementations are freely available and they are supported on most major computers and operating systems. Students can install the software on their own machines, as well as using the university's computers.

The Software Tools for Discrete Mathematics package is a library of definitions that are loaded into Haskell. This package is available on the book web page (see Appendix B).

Haskell is the ideal language for teaching discrete mathematics. It offers a powerful and concise expression language; many problems that would require writing a complete program on the order of 10 to 100 lines of code in a language such as Pascal, C++ or Java can be written as a simple expression in Haskell which is only a few lines long. This makes it possible to use Haskell interactively to experiment with the mathematical expressions of propositional logic, predicate logic, set theory and relations. Such lightweight interactive exploration is infeasible in traditional imperative or object-oriented languages. Haskell is also well suited for complex applications, such as the proof checker used in Chapters 2 and 3, and the hardware description language used in Chapter 10.

It is assumed that the reader of the book has no knowledge in advance about Haskell or functional programming; everything that is needed is covered here. Since it is self-contained, this book can be used in any curriculum, regardless of what programming languages happen to be in use.

To the Student

It's best to read this book *actively* with pencil and paper at hand. As you read, try out the examples yourself. It is especially important to try some of the exercises, and solutions to many of them appear in Appendix C. Don't just read the exercise and then the solution—the benefit comes from trying to solve an exercise yourself, even if you don't get it right. When you find your own solution, or if you get stuck, then compare your solution with the one in the book.

The web page for this book has additional information that will be useful as you study discrete mathematics:

```
http://www.dcs.gla.ac.uk/
     ~jtod/discrete-mathematics/
```

Many of the exercises require the use of a computer with Haskell installed. The software is free, and it's straightforward to download it and install on your own machine. See the book web page for information on obtaining the software.

A good way to improve your understanding of the material is to read about it at a more advanced level, and also to learn about its application to real problems. The Bibliography near the end of the book lists many good sources of information, and each chapter ends with some suggestions for further reading.

We wish you success with your studies in mathematics and computer science!

To the Instructor

This book is primarily intended for students of computer science, and applications of the mathematics to computing are stressed. No specific topics in computing are prerequisites, but some familiarity with élementary computer programming is assumed. The level is appropriate for courses in the first or second year of study. The contents of this book can be covered in a course of one semester.

The *Instructor's Guide* gives suggestions for organising the course, solutions to the exercises, additional problems with solutions and other teaching resources. It is available online:

```
http://www.dcs.gla.ac.uk/~jtod
    /discrete-mathematics/instructors-guide/
```

Notation

Standard mathematical notation is used in this book when discussing mathematics: $A \subseteq B$. A typewriter font is used for notations that are intended to be input to a computer: a 'subset' b. For example, a general discussion in English might say that a theorem is true; that theorem might make a statement about the proposition True, and a Haskell program would use the constant True. The end of a proof is marked by a square box □.

Acknowledgements

We would like to thank the following colleagues for their helpful feedback and encouragement during the process of writing this book: Tony Davie, Bill Findlay, Joy Goodman, Mark Harman, Greg Michaelson, Genesio Gomes da Cruz Neto, Thomas Rauber, Richard Reid, Gudula Rünger and Noel Winstanley. We would also like to thank the students at the University of Glasgow and the University of Michigan who gave both of us experience teaching with preliminary versions of this material, and our editor, Karen Barker, for her help in producing this book. All remaining errors are ours alone.

Cordelia Hall and John O'Donnell
Glasgow
September, 1999

Contents

Chapter 1

Introduction to Haskell

The topic of this book is discrete mathematics with an emphasis on its connections with computers:

- *The computer can help you to learn and understand mathematics.* As various mathematical objects are defined, software tools will enable you to perform calculations with those objects. Exploring and experimenting with mathematical ideas gives a practical intuition.

- *The mathematics has widespread applications in computing.* We will focus on the topics of discrete mathematics that are most important in modern computing, and will look at many examples.

- *Software tools make it possible to use the mathematics more effectively.* Mathematical structures are frequently large and complex, and computers are often necessary to bring out their full potential.

In order to achieve these goals, it won't be enough to provide the occasional pseudo-code program. We need to work with real programs throughout the book. The programming language that will be used is Haskell, which is a modern standard functional programming language.

It is *not* assumed that you know anything about Haskell, or about functional programming. Everything you need to know about it is covered in this book. You will need only a small part of the language for this book, and that part is introduced in this chapter. If you would like to learn more about Haskell or functional programming, Section 1.11 recommends a number of sources for further reading.

Why use a functional language, and why Haskell in particular? Because:

- Haskell allows you to compute directly with the fundamental objects of discrete mathematics.

- It is a powerful language, allowing programs that would be long and complicated in other languages to be expressed simply and concisely.

1

- You can reason mathematically about Haskell programs in the same way you do in elementary algebra.

- The language provides a strong type system that allows the compiler to catch a large fraction of errors; it is rare for a Haskell program to crash.

- Haskell is an excellent language for *rapid prototyping* (i.e. implementing a program quickly and with minimal effort in order to experiment with it).

- There is a stable, standard and well-documented definition of Haskell.

- A variety of implementations are available which are free and which run on most computers and operating systems.

- Haskell can be used interactively, like a calculator; you don't need a heavy-weight compiler.

1.1   Obtaining and Running Haskell

Appendix A gives a pointer to the web page for this book. On that page you can find pointers to the most up-to-date implementations of Haskell as well as the software tools used in this book. There is an active Haskell development community, and new tools are constantly emerging, so you should check the book web page for current information. The book web page also contains additional documentation for the software.

We will use the computer interactively, like a desk calculator. Haskell itself provides a powerful set of built-in operations, but others that you will need are defined in the *Software Tools for Discrete Mathematics* file `stdm.hs` which you can download from the book web page.

To give an idea of what it's like to use the computer with this book, here is a typical interactive session, using the `stdm` file and the Hugs98 implementation.

First we start the Hugs98 program, and it will give an introductory screen followed by a prompt '>'. Haskell is now acting like an interactive calculator; you enter an expression after the > prompt, and it will evaluate the expression, print it and give another prompt:

```
... the introductory message from Haskell system ...
Type :? for help
> 1 + 2
3
> 3*4
12
>
```

You have just typed in an *expression*, and it was evaluated. An expression is a combination of operators and values that are defined by the programming language. For example, 1 + 2 * x is an expression in Haskell. We write 1+2 $\Rightarrow$ 3, meaning that when the expression 1+2 is evaluated, the result is 3.

Now you should load the software tools file, which defines the programs used in the examples and exercises in this book:

```
> :load stdm
```

Notice the colon in the :load stdm. If the first character after a prompt is : then the rest of the line is a command to the interpreter, rather than an expression. One good command to remember is :?, which prints a help screen listing all the other commands you can enter.

At this point, you are ready to start writing and testing definitions. These should be saved in a file, so that you don't have to enter them in again and again. To do this, create a file with a name like mydefs.hs. The extension .hs stands for *Haskell script*. Store the following two lines in your file:

```
y = x+1
x = 2*3
```

The equations in your script file really are mathematical equations; they aren't assignment statements. One consequence of this is that you can write the equations in whatever order seems easiest to understand; there is no need, for example, to put the equation defining x before the equation defining y. Another consequence is that the same method of 'substituting equals for equals' used in mathematics can also be applied to Haskell. For example, the equation says y = x+1, and we have another equation that says x = 2*3; substituting 2*3 for x yields y = (2*3)+1. This ease of reasoning mathematically about Haskell programs is one of the central reasons that Haskell is such a suitable language in which to write mathematical software.

Now enter the command :load mydefs, which tells Hugs to read your file. The effect of this is to define all the names that appear on the left hand sides of the equations in the file. From now on, the expressions you enter interactively can make use of the variables x and y:

```
:load mydefs
> :load mydefs
Reading file "mydefs.hs":
mydefs.hs
> x
6
> y
7
> x*y
42
>
```

You can always edit the file and save it, and then reload the file into Haskell. This will replace all the old definitions with the ones in the edited file. For example, if you modify the file so that it says x = 4*5 and reload it, then x ⇒ 20 and y ⇒ 21.

When you would like to leave Hugs, you can enter the quit command :quit, and with some operating systems you can also quit by entering control D.

1.2 Expressions

You can do a lot just by writing simple expressions in Haskell—far more than just basic arithmetic (although that can be interesting, too). In the following sections, we will show some of the most useful kinds of expression, organised according to the *type* of value.

As you will see later, types are of fundamental importance, both in Haskell and in discrete mathematics. For now, however, you can just think of a type as being something like Integer, Float, Char, etc. The essential point is that an operator is defined for a specific type. For example, the + operator specifies addition, which makes sense only for number types, while the length operation gives the length of a list, but makes no sense for other types.

1.2.1 Integer and Int

Integer constants are written as a sequence of digits. For example, 2, 0, 12345 and -72 are all integer constants.

Like most programming languages, Haskell provides operators for addition (+), subtraction (-) and multiplication (*). There is also an exponentiation operator ^; for example, 2^3 means '2 raised to the power 3', or 2^3.

For the time being, don't use / for division; instead, write x 'div' y in order to divide x by y. This is an integer division, and any remainder is thrown away. For example, 5 'div' 2 ⇒ 2, 17 'div' 3 ⇒5 and -100 'div' 20 ⇒ -5. You can get the remainder or modulus with the 'mod' operator: 8 'mod' 3 ⇒ 2.

Notice how 'div' and 'mod' are *operators* (an operator is written in between its two arguments), but these operators are names made up of letters. Haskell has many named operators like this, and they are always enclosed in back-quote characters: 'opname'.

There are functions that find the larger and smaller of two numbers: max 3 8 ⇒ 8, and min 3 8 ⇒ 3. There is a further discussion about operators like 'div' and functions like max in Section 1.4.3.

Computers store numbers in *words*, and the word size of modern processors is 64 bits long. Whatever the word size on your computer, however, there is a limit to how large an integer it will hold. Fortunately, Haskell does not limit you to numbers that fit into a word. It provides two distinct integer types:

- `Int` is the type of integers that fit into a word on the computer (naturally this will vary from one computer to another);

- `Integer` is the type of mathematical integers.

Since there are two different integer types, we need a way of saying which type we want an expression to have. The `::` operator (read it as *has type*) is used in Haskell to specify what type an expression has. Thus `2::Int` says '2 has type `Int`', and its representation in the computer is different from `2::Integer`.

Here is an example that will illustrate the difference between `Int` and `Integer`. First, evaluate `2^2`, which means 2 squared and gives 4:

```
> 2^2
4
```

Now, `2^20` presents no problem to most computers, as the word size will be well above 20:

```
> 2^20
1048576
```

However, most computers have a word size much less than 200 bits, and the expression `2^200` will not give the right answer if it's evaluated with the default `Int` type:

```
> 2^200
0
```

Therefore we say explicitly that we want the unlimited `Integer` type to be used instead; now we are guaranteed to get the right answer:

```
> (2^200)::Integer
1606938044258990275541962092341162602522202993782792835301376
```

One might wonder just how large an `Integer` number can be. When it performs arithmetic on these numbers, Haskell allocates enough memory dynamically in order to store the result. There is of course a limit to the size of number that can be stored, but a modern machine with a large memory can easily accommodate numbers that contain millions of digits.

Besides actually needing large numbers, there is a theoretical benefit from using `Integer`: the arithmetic operations on this type satisfy the algebraic laws. For example, we know from algebra that $(x + y) - y = x$, but this is *not* always true in a computer, if the arithmetic is performed with the `Int` type, or in a language that offers only fixed-word arithmetic. It might happen that x and y fit in a machine word, but the intermediate result $x + y$ does not. In contrast, if the arithmetic is performed on the `Integer` type then the computer program will definitely satisfy the mathematical law.

1.2.2   Rational and Floating Point Numbers

Single-precision floating point numbers have type `Float`, and double-precision numbers have type `Double`. Besides the operators `+`, `-` and `*`, you can divide floating point numbers with the `/` operator. The floating point exponentiation operation is `**`. For example, `2**0.5` $\Rightarrow$ `1.41421`.

There are also a number of functions that can be applied to floating point numbers. A *function application* in Haskell requires no parentheses; you just write the name of the function, followed by a space, followed by the argument. For example, the square root function is `sqrt`, and a typical application is `sqrt 9` $\Rightarrow$ `3.0`.

Floating point representations are approximations, and they are *not* guaranteed to satisfy the algebraic laws as `Integer` numbers do. Try evaluating the following expressions on your computer:

```
0.11 - 0.10
2.11 - 2.10
```

If the arithmetic were performed exactly, these would both give the same result, but they do not on some computers. This is a property of floating point representation, not Haskell. It is possible to round the numbers so that they look the same when printed, but the internal representations will still be different. It is important to remember that when you use mathematics to reason about real numbers, the results may not apply exactly to a program that uses floating point. It is particularly important to be careful when comparing floating point numbers: the right way to compare them is to determine whether the absolute value of their difference falls within an acceptable error tolerance.

Haskell supports *exact* arithmetic on rational numbers, allowing you to work with fractions as well as with their decimal equivalents (approximations). `Ratio Integer` is the type of rational numbers; these are numbers in the form of a fraction, where the numerator and denominator are represented with the `Integer` type. A rational constant is written in the form *numerator%denominator*. You can divide two integers using the `/` operator, and the result is an exact rational number. Haskell automatically reduces fractions; for example:

```
2/3                           ⇒   0.66667 ::  Double
2/3 :: Ratio Integer          ⇒   2%3 ::  Ratio Integer
2%3 + 1%6                     ⇒   5%6 ::  Ratio Int
(1/3 + 1/4) :: Ratio Integer  ⇒   7%12
```

1.2.3   Booleans

The `Bool` type is used to represent the result of comparisons, and similar operations, where the result must be either `True` or `False`. These are the only two values with type `Bool`.

The following operators can be used to compare two numbers, and they produce a result of type `Bool`:

```
==  -- equality
/=  -- not equal
<   -- less than
<=  -- less than or equal
>   -- greater than
>=  -- greater than or equal
```

For example, 9>3 ⇒True, and 5<=5 ⇒True.

There are also some operators that take arguments of type `Bool`, again returning a `Bool` result:

```
&&  -- Boolean and
||  -- Boolean or
not -- Boolean not
```

The expression x&&y evaluates to `True` if *both* x and y are `True`; x||y evaluates to `True` if *either* of the arguments is `True`. Finally, not x is `True` if x is `False`, and vice versa.

Exercise 1. Evaluate these expressions, and check with the computer:

```
True && False
True || False
not False
3 <= 5 && 5 <= 10
3 <= 20 && 20 <= 10
False == True
1 == 1
1 /= 2
1 /= 1
```

1.2.4 Characters

A character has type `Char`, and a constant is written as the character surrounded by single-quote characters; for example, `'a'` and `'*'`. Recall that the back-quote character is used for operators, *not* for characters. Thus `'?'` is a character and `` `div` `` is an operator.

The `:type` command in Hugs will tell you the type of an expression. As you experiment with the language by evaluating expressions, it is also a good idea to use `:type` to be sure that they have the same type you think they should. Often you will see a type with => in it; the meaning of this is explained in Section 1.10.

The comparison operators can be used on characters. Two useful built-in functions are `toUpper` and `toLower`, which take a character and convert it from one case to another. For example:

```
'c' < 'Z'
toUpper 'w' => 'W'
toLower 'Q' => 'q'
```

There is a special character called *newline* which causes a line break when it is printed. This character is written '\n'.

1.2.5 Strings

A `String` is a sequence of zero or more characters. A string constant is written inside double-quote marks. For example, "tree"::String.

There is a difference between 'a'::Char and "a"::String. The first is the character a, while the second is a string that *contains* the character a. A string can have any length, unlike a character.

The `length` function can be used to determine how many characters are in a string. For example, length "dog" ⇒ 3, and length "a" ⇒ 1. You cannot apply length to a character.

Two strings can be *concatenated* together with the operator ++ to form a larger string: "abc" ++ "defg" ⇒ "abcdefg". A common example is to place a newline character at the end of a line: "Here is a line" ++ "\n".

The `length` function and ++ operator are actually more general than described here; this will be discussed later.

1.3 Basic Data Structures: Tuples and Lists

The most commonly used data structures in Haskell are tuples and lists. Both of them are used to gather several data values together into one data structure.

1.3.1 Tuples

Tuples package up several values into one. They are written as a sequence of data values, separated by commas, inside parentheses. For example, (2,"dog") is a tuple with two components; the first is 2 and the second is "dog". Its type is written (Int,"dog"). In general, if x::A and y::B, then (x,y)::(A,B). For example:

```
("dog","cat") :: (String,String)
(True, 5) :: (Bool,Int)
('a',"b") :: (Char,String)
("bat",(3.14,False)) :: (String,(Double,Bool))
```

The usual way to extract the components from a tuple is pattern matching, which will be covered later. For tuples with 2 components, the function fst returns the first component and snd returns the second. These functions have the following types:

```
fst :: (a,b) -> a
snd :: (a,b) -> b
```

This says that `fst` is a function; its argument has a tuple type of the form
(`a`,`b`), and the result it returns has type `a`.

A tuple can have any number of components. If it contains n components, it
is called an n-tuple, but 2-tuples are often just called pairs. You can even have
a 0-tuple which contains nothing, written (); this is often used as a dummy
value. (But you cannot have a 1-tuple, as that would conflict with the use of
parentheses to control precedence of operators.)

Tuples have two important characteristics that distinguish them from lists.
First, a tuple has a fixed number of components. If you have a 3-tuple (`a`,`b`,`c`),
and you add an extra data value to obtain (`a`,`b`,`c`,`d`), the new tuple has a
different type from the old one. Second, there is no restriction on the type of
any component: it is common for a tuple to contain data values with different
types.

1.3.2   Lists

Lists are the most commonly used data structure in functional languages. A
list is written as a sequence of elements, separated by commas, surrounded by
square brackets: for example, [1,2,3] is a list of integers.

A list may contain any number of elements, but all of the elements *must*
have the same type. The type of a list is written [`A`], where `A` is the element
type. For example:

```
[13,9,-2,100] :: [Int]
["cat","dog"] :: [String]
[[1,2],[3,7,1],[],[900]] :: [[Int]]
```

Expressions can appear in a list expression; all the elements of a list are evaluated. If your script file contains an equation defining x, then [13,2+2,5*x]
is a list as well.

A `String` is actually just a list of characters. The constant "abc" is exactly
equivalent to writing ['a','b','c'].

You can easily specify a list which is a sequence of numbers. For example,
the Haskell notation [1..10] means the list [1,2,3,4,5,6,7,8,9,10], and
the .. is filled in with the missing numbers. Normally the numbers are incremented by 1, but if you give two numbers before the .. any increment can be
used. For example, [1,3..10] counts up by 2, and its value is [1,3,5,7,9].
Sequences of characters can be created in the same way. Here are some examples:

```
['a'..'z'] => "abcdefghijklmnopqrstuvwxyz" :: String
['0'..'9'] => "0123456789" :: String
[0..9]     => [0,1,2,3,4,5,6,7,8,9] :: [Int]
[10,9..0]  => [10,9,8,7,6,5,4,3,2,1,0] :: [Int]
```

Haskell has a large number of features to make lists easy to use. Some of these are presented in Section 1.8.

1.3.3 List Notation and (:)

A new element can be added to the front of a list using the operator (:), which is usually pronounced 'cons' (because it constructs a list). This function takes a value and a list and inserts the value at the front of the list:

```
(:) :: a -> [a] -> [a]

1:[2,3] => [1,2,3]
1:[] => [1]
```

Every list is built up with the (:) operator, starting from the empty list []. Thus 1:[] is a list containing one element, the number 1. This list can be written just as [1]; in fact, the usual notation for lists is merely a nicer syntax for an expression that builds the list. This notational convention allows the following equations to be written:

```
[1,2,3,4] = 1 : (2 : (3 : (4 : [])))
"abc" = 'a' : ('b' : ('c' : []))
```

The parentheses can be omitted, so the equations could be written more simply:

```
[1,2,3,4] = 1 : 2 : 3 : 4 : []
"abc" = 'a' : 'b' : 'c' : []
```

1.3.4 List Comprehensions

The list structure has been so useful and influential in functional languages that Haskell offers a variety of features to make them convenient to use. The *list comprehension* is a simple but powerful syntax that lets you define many lists directly, without needing to write a program to build them. List comprehensions are based on the standard set comprehension notation in mathematics. For example, the set comprehension $\{x^2 \mid x \in S\}$ describes the set of squares of elements of another set S.

The basic form of list comprehension is a list in which an expression appears first, followed by a vertical bar (read 'such that'), followed by a *generator*:

```
[expression | generator]
```

The generator specifies a sequence of values that a variable takes on; this is written in the form var <- list, and it means that the variable var will take on each of the values in list, one by one. For each of those values, the expression to the left of the bar is evaluated, and the result goes into the list.

For example, the following list comprehension says that x should take on the values 1, 2 and 3; for each of these the value of 10*x goes into the result list, which is [10,20,30]:

```
[10*x | x <- [1,2,3]]
   => [10,20,30]
```

Often a sequence with the .. notation is used to define the list from which the variable draws its values. For example:

```
[x^2 | x <- [1..5]]
   => [1,4,9,16,25]
[y 'mod' 3 | y <- [1..6]]
   => [1,2,0,1,2,0]
[toLower c | c <- "Too Many CAPITALs"]
   => "too many capitals"
```

If the list to the right of the <- is actually a list of tuples, then the variable can be replaced by a tuple of variables. In the following example, the tuple (a,b) takes on the values (1,2), (10,20) and (6,6). For each of these list elements, the expression a*b multiplies the components of the tuple:

```
[a*b | (a,b) <- [(1,2), (10,20), (6,6)]]
   => [2,200,36]
```

List comprehensions can have more than one generator. For each value of the first generator taken by the first variable, the second variable gets all of the values in its generator. This is similar to a nested loop, where the first generator is the outer loop and the second generator is the inner one. For example,

```
[(x,y) | x <- [1,2,3], y <- ['a','b']]
   => [(1,'a'),(1,'b'),(2,'a'),(2,'b'),(3,'a'),(3,'b')]
```

In the examples so far, all of the elements of the generator list cause a corresponding value to appear in the result. Sometimes we want to throw out some of the elements of the generator. This is performed by using a *filter*. A filter is an expression of type Bool, which (normally) uses the generator variable. After the variable takes on a value of the generator, the filter is evaluated, and if it is False then the value is thrown out. For example, the following list comprehension makes a list of all the integers from 0 to 100 which are multiples of both 2 and 7:

```
[x | x <- [0..100], x 'mod' 2 == 0 && x 'mod' 7 == 0]
   => [0,14,28,42,56,70,84,98]
```

The next comprehension produces a list of all of the factors of 12. It works by considering all combinations of x and y between 0 and 12; the value of x is retained if the product is exactly 12. If you replace 12 with any other number, it will make a list of the factors of that number:

```
[x | x <- [1..12], y <- [1..12], 12 == x*y]
```

List comprehensions provide an easy way to implement database queries. For example, db is a database represented as a list of tuples. Each tuple contains a person's name, their street address, age, and annual income:

```
db = [("Ann Smith",  "29 Byres Road",  30, 48000),
      ("Alan Jones", "36 High Street", 25, 17000),
        ...
     ]
```

Now we can use a list comprehension to make a mailing label for every employee under 30 who is making at least £15000 per year:

```
[name ++ "\n" ++ addr ++ "\n"
  | (name,addr,age,sal) <- db, age<30, sal>=15000]
```

Exercise 2. Work out the values of the following list comprehensions; then check your results by evaluating them with the computer:

```
[x | x <- [1,2,3], False]

[not (x && y) | x <- [False, True],
                y <- [False, True]]

[x || y |  x <- [False, True],
           y <- [False, True],
           x /= y]

[[x,y,z] | x <- [1..50], y <- [1..50], z <- [1..50],
           x ** 2 + y ** 2 == z ** 2]
```

1.4   Functions

Functions are, unsurprisingly, central to functional programming languages like Haskell. We have already seen a few brief examples. Now we take a closer look at functions: how to use them, their types, and how to define them.

1.4.1 Function Application

An expression using a function to perform a computation is called a *function application*, and we say that a function is *applied* to its arguments. For example, we might apply the function `sqrt` to the argument 9, obtaining the result 3.0.

The notation for a function application consists of the function name, followed by a space, followed by the function argument(s). If there are several arguments, they should be separated by spaces. Unlike some other languages, you do not put parentheses around the arguments of a function.

```
sqrt 9.0
3.4 + sqrt 25 * 100
2 * sqrt (pi * 5 * 5) + 10
```

1.4.2 Function Types

Just as data values have types, so do functions. Function types are important because they say what type of argument a function requires, and what type of result it will return.

A function that takes an argument of type a and returns a result of type b has a *function type* which is written $a \rightarrow b$ (read a arrow b). To say that f has this type, we write $f :: a \rightarrow b$. In Haskell, the $\rightarrow$ symbol is written as `->`. For example, some of the Haskell functions that we have already seen have the following types:

```
sqrt :: Double -> Double
max :: Integer -> Integer -> Integer
not :: Bool -> Bool
toUpper :: Char -> Char
```

1.4.3 Operators and Functions

An *operator* is a function in Haskell, but an operator must take exactly two arguments, and it is written between the arguments. For example, the + operator takes two numbers and adds them, and is written between the numbers: 2+3. Since an operator is a function, it has a function type. When you use the operator by itself, for example to state its type, it must be enclosed in parentheses:

```
(+) :: Integer -> Integer -> Integer
(&&) :: Bool -> Bool -> Bool
```

Since an operator is just a function, you can do everything with an operator that you can do with a function, but it must be enclosed in parentheses if it does not appear between its arguments. For example, an alternative way to write 2+3 is (+) 2 3, which says that the (+) function is applied to two arguments,

2 and 3. Later in this book you will see a number of examples where it is useful to do this.

Just as you can treat an operator as a function by putting parentheses around it, you can treat a function as an operator by putting single-quote characters around it. For example, the function `max` can be applied to two arguments in a function application: `max 4 7`. If you prefer, you can use it as an operator, like this: `4 'max' 7`.

1.4.4 Function Definitions

You can define new functions by giving the *type declaration* followed by the *defining equation*. The type declaration has the form:

$$function\_name :: argType_1 \rightarrow argType_2 \rightarrow \ldots \rightarrow argType_n \rightarrow resultType$$

The arrows are written as $\rightarrow$ in mathematical notation, and they are written as `->` in Haskell programs. The defining equation has the form:

$$function\_name\ arg_1\ arg_2\ \ldots\ arg_n = expression\ that\ can\ use\ the\ arguments$$

The function definition should be inserted in a Haskell script file. When you load the file, you will then be able to use your functions. For example, suppose we want a function `square` that takes an `Integer` and squares it. Here is the function definition:

```
square :: Integer -> Integer
square x = x*x
```

If you just evaluate the expression `x` at the top level, Haskell will give an error message because `x` is not defined:

```
> x
ERROR: Undefined variable "x"
```

However, when a function is applied to an argument, the value of the argument is available to the expression on the right hand side of the function's defining equation. For example, if we evaluate `square 5`, then `x` is temporarily defined to be 5, and the right hand side `x*x` is evaluated, producing the result 25.

1.4.5 Pattern Matching

As we have seen, the left hand side of the defining equation has the form of an application. If the argument in that application is a name (for example `x` in the defining equation `square x = x*x`) then when an application is evaluated, the name will take on the argument value, and the right hand side of the function is evaluated.

There is another option: if the argument on the left hand side is a constant, then the right hand side will be used only if the function is applied to that value. This makes it possible for a function definition to consist of several defining equations. For example, here is a definition of f with three defining equations, each with a constant argument:

```
f :: Integer -> String
f 1 = "one"
f 2 = "two"
f 3 = "three"
```

When the application f 2 is evaluated, the computer begins by checking the first defining equation, and it discovers that the application f 2 does not match the left hand side of the first line, f 1. Therefore the next equation is tried; this one does match, so the corresponding right hand side is evaluated and returned. If f is applied to an argument like 4, which does not match any of the defining equations, an error message is printed and the program is terminated.

The following function tests its value, returning True if the value is a 3, and False otherwise. It has two defining equations. If the argument pattern 3 in the first defining equation matches the argument, then that equation is used. If the argument is not 3, then the second equation is checked; the argument x will match *any* argument, so the corresponding value False is returned.

```
is_three :: Int -> Bool
is_three 3 = True
is_three x = False
```

For example, given the application is_three (1+1), the argument is evaluated to 2; since this does not match 3, the second defining equation is used.

A function may have more than one argument. If this is the case, then each argument in a given defining equation is checked, from left to right. For example, here is the nor function, which returns True if and only if both of the arguments are False:

```
nor :: Bool -> Bool -> Bool
nor False False = True
nor a     b     = False
```

Consider the evaluation of the application nor False True. The first defining equation is checked from left to right at runtime. The first pattern matches, so the second argument on that line is also checked. It does not match, so the second defining equation is checked; its left hand side matches both arguments.

Pattern matching is commonly used when the argument to a function is a tuple. For example, suppose we want a function fst which takes a pair and returns its first element, and a similar function snd to return the second element. These functions are easily defined with pattern matching:

```
fst :: (a,b) -> a
fst (x,y) = x

snd :: (a,b) -> b
snd (x,y) = y
```

When `fst (5,6)` is evaluated, the argument value `(5,6)` is matched against the argument pattern `(x,y)` in the defining equation; this causes `x` to be defined as 5 and `y` as 6. Then the right hand side `x` is returned, giving the result 5. The `fst` and `snd` functions are very useful, so they are already defined in the standard Haskell library.

An argument pattern may also be a list. Since a list is either `[]` or else it is constructed with the cons operator `:`, the patterns take the forms `[]` and `x:xs`. For example, the following function determines whether a list is empty:

```
isEmpty :: [a] -> Bool
isEmpty [] = True
isEmpty (x:xs) = False
```

The parentheses around the pattern `x:xs` in the left hand side are required because otherwise the compiler would parse it incorrectly.

When `isEmpty []` is evaluated, the argument matches the pattern in the first defining equation, so `True` is returned. However, when the application `isEmpty (1:2:3:[])` is evaluated, the argument fails to match `[]` and the second equation is tried. Here the match succeeds, with `x` defined as 1 and `xs` defined as `2:3:[]`, and `False` is returned. It makes no difference if the argument is written with the simpler syntax `[1,2,3]`; since this is merely an abbreviation for `1:2:3:[]`, the evaluation is identical.

The following functions, which are defined in the standard Haskell libraries, can be used to return the first element of a list (the head), or everything *except* the first element (the tail):

```
head :: [a] -> a
head (x:xs) = x

tail :: [a] -> [a]
tail (x:xs) = xs
```

Exercise 3. Write a function that takes a character and returns True if the character is 'a' and False otherwise.

Exercise 4. Write a function that takes a string and returns True if the string is `"hello"` and False otherwise. This can be done by specifying each element of the string in the list pattern (e.g. `'h':'i':[]`).

Exercise 5. Write a function that takes a string and removes a leading space if it exists.

1.4.6 Equational Reasoning

The equations that appear in a Haskell script are genuine mathematical equations, stating that the left and right hand sides have the same value. They are not commands, and they do not cause the value of a variable to change. Imperative programming languages are based on the assignment statement; thus n := x*y means that x*y is evaluated, and the result is stored in the variable n, erasing its previous value. For example, the imperative assignment statement n := n+1 increases the value of n, while the Haskell equation n = n+1 defines n to be the solution to the mathematical equation. Since this equation has no solution, the result is that n is undefined. If the program ever uses an undefined value like n, the computer will either give an error message, or else go into an infinite loop.

The fact that Haskell uses only equations leads to a powerful technique for reasoning about functional programs, called *equational reasoning*. This is the same kind of reasoning used in algebra, often called 'substituting equals for equals'. For example, suppose we have the following script file:

```
a = 3
b = 4
c = 5
z = a^2 + b^2 + c^2

f :: Integer -> Integer -> Integer
f x y = 2*x + y
```

Now the expression f a (f z 2) can be evaluated by equational reasoning. We use the convention in this book that Haskell definitions are in typewriter font while mathematical reasoning is in *mathematical italic notation*.

$$
\begin{aligned}
f\, a\, & (f\, z\, 2) \\
&= f\, a\, (f\, (a*a + b*b + c*c)\, 2) \\
&= f\, a\, (f\, (3*3 + 4*4 + 5*5)\, 2) \\
&= f\, a\, (f\, 50\, 2) \\
&= f\, a\, (2*50 + 2) \\
&= f\, a\, 102 \\
&= f\, 3\, 102 \\
&= 2*3 + 102 \\
&= 108
\end{aligned}
$$

At each step, the previous expression is rewritten using ordinary mathematical laws and the Haskell definitions, all of which are equations. Equations are timeless, since there is no notion of changing the value of a variable. If assignment statements were part of the language, this simplification technique could not be used.

1.4.7   Higher Order Functions

Functions in Haskell are 'first class objects'; that is, you can store them in data structures, pass them as arguments to functions, and create new ones. A function is called *first order* if its arguments and results are ordinary data values, and it is called *higher order* if it takes another function as an argument, or if it returns a function as its result. Higher order functions make possible a variety of powerful programming techniques.

The `twice` function takes another function `f` as its first argument, and it applies `f` two times to its second argument `x`:

```
twice :: (a->a) -> a -> a
twice f x = f (f x)
```

We can work out an application using equational reasoning. For example, `twice sqrt 81` is evaluated as follows:

$$
\begin{aligned}
&twice\ sqrt\ 81 \\
&= \ sqrt\ (sqrt\ 81) \\
&= \ sqrt\ 9 \\
&= \ 3
\end{aligned}
$$

Let's examine the type of `twice` in detail. Assuming the second argument has type `a`, then the argument function has to accept an argument of type `a` and also return a value of the same type (because the result of the inner application becomes the argument to the outer application). Hence the function argument must have type `a->a`.

In Haskell, functions receive their arguments one at a time. That is, if a function takes two arguments but you apply it to just one argument, the evaluation gives a new function which is ready to take the second argument and finish the computation. For example, the following function takes two arguments and returns their product:

```
prod :: Integer -> Integer -> Integer
prod x y = x*y
```

A *full application* is an expression giving `prod` all of its arguments: for example, `prod 4 5` ⇒ 20. However, the *partial application* `prod 4` supplies just one argument, and the result of this is a new function which takes a number and multiplies it by 4. For example, suppose the following equations are defined in the script file:

```
g = prod 4
p = g 6
q = twice g 3
```

Then p ⇒ 24, and q ⇒ 48.

1.5 Conditional Expressions

A conditional expression uses a `Bool` value to make a choice. Its general form is:

if *boolean_expression* then *exp1* else *exp2*.

The boolean expression is first evaluated; if it is `True` then the entire conditional expression has the value of *exp1*; if it is `False` the whole expression has the value of *exp2*.

A conditional expression must always have both a `then` expression and an `else` expression. Both of these expressions must have the same type, which is the type of the entire conditional expression. For example,

```
if 2<3 then "bird" else "fish"
```

has type `String` and its value is `"bird"`. However, the following expressions are incorrect:

```
if 2<3 then 10              -- no else expression
if 2+2 then 1 else 2        -- must be Bool after if
if True then "bird" else 7  -- different types
```

The `abs` function returns the absolute value of its argument, and it uses a conditional expression since the result depends on the sign of the argument:

```
abs :: Integer -> Integer
abs x = if x<0 then -x else x
```

1.6 Local Variables: `let` Expressions

There are many times when we need to use computed values more than once. Instead of repeating the expression several times, it is better to give it a local name which can be reused. This can be done with a `let` expression. The general form is:

let *equation*

equation

⋮

equation

in *expression*

This entire construct is just one big expression, and it can be used anywhere an expression would be valid. When it is evaluated, the local equations give temporary values to the variables in their left hand sides; the final expression after `in` is the value of the entire `let` expression. Each of the local definitions can refer to the others.

For example, consider the following function definition which defines the two real solutions (x_1, x_2) of the quadratic formula $a \times x^2 + b \times x + c = 0$:

```
quadratic :: Double -> Double -> Double -> (Double,Double)
quadratic a b c
  = let d = sqrt (b^2 - 4*a*c)
        x1 = (-b + d) / (2*a)
        x2 = (-b - d) / (2*a)
    in (x1,x2)
```

The `let` expression gives the three variables `d`, `x1` and `x2` values which can be used locally. For example, the value of `d` is used in the equations for `x1` and `x2`. Outside of this `let` expression, the names `d`, `x1` and `x2` are not defined by these equations (although they might be defined by equations in some enclosing `let` expression).

Let expressions are *expressions*! For example, this is not a block of statements that gets executed; it is an expression that uses a local definition:

```
2 + let x = sqrt 9 in (x+1)*(x-1)
```

1.7   Type Variables

Recall the functions `fst` and `snd`, which return the first and second component of a pair. There is actually an infinite number of different types these functions could be applied to, including:

```
(Integer,Float)
([Char], [Integer])
(Double, (Int,Char))
```

It would not be good to give `fst` a restrictive type, like `(Integer,Float)` `-> Integer`; this would work for the first example above, but not the others. We really want to say that `fst` takes a pair whose components may have *any* type, and its result has the same type as the first component. This is done by using *type variables*:

```
fst :: (a,b) -> a
snd :: (a,b) -> b
```

Type variables must begin with a lower case letter, and it is a common convention to use `a`, `b` and so on.

A function with a type variable in its type signature is said to be *polymorphic*, because its type can take many forms. Many important Haskell functions are polymorphic, enabling them to be used in a wide variety of circumstances. Polymorphism is a very important invention because it makes it easier to reuse programs.

1.8 Common Functions on Lists

Haskell provides a number of operations on lists. A few of the most important ones are presented in this section. We will have more to say about these functions later in the book, where we will use them in a variety of practical applications, show how they are implemented, and prove theorems stating some of their mathematical properties.

The length function

The length function returns the number of elements in a list:

```
length :: [a] -> Int

length [2,8,1] => 3
length [] => 0
length "hello" => 5
length [1..n] => n
length [1..] => <infinite loop>
```

The !! (index) operator

The (!!) operator lets you access a list element by its index. Indices start from 0.

```
(!!) :: [a] -> Int -> a

[1,2,3] !! 0 => 1
"abcde" !! 2 => 'c'
```

The take function

This function takes the first n elements from a list:

```
take :: Int -> [a] -> [a]

take 2 [1,2,3] => [1,2]
take 0 [1,2,3] => []
take 4 [1,2,3] => [1,2,3]
```

The drop function

The drop function drops the first n elements from a list; it removes exactly the same elements that take would return:

```
drop :: Int -> [a] -> [a]

drop 2 [1,2,3] => [3]
drop 0 [1,2,3] => [1,2,3]
drop 4 [1,2,3] => []
```

The ++ (append) operator

Two lists can be joined together using the append (also called concatenation)
(++) operation. All of the elements in the resulting list must have the same
type, so the two lists must also have the same type.

```
(++) :: [a] -> [a] -> [a]

[1,2] ++ [3,4,5] => [1,2,3,4,5]
[] ++ "abc" => "abc"
```

The map function

Often, we want to apply a function to each element in a list. For example, we
may have a string of lower case letters and want them to be upper case letters.
To do this, we use map, which takes a function of one argument and a list. It
applies the function to each element of the list.

```
map :: (a -> b) -> [a] -> [b]

map toUpper "the cat and dog" => "THE CAT AND DOG"
map (** 2) [1,2,3] => [1,4,9]
```

Exercise 6. Write a function that uses map and takes a list of Ints and converts
them into Boolean (e.g. 0 becomes False and 1 becomes True).

Exercise 7. The function or takes a list of Boolean values and returns True
if at least one of them is True and False otherwise. Write a function that
uses map and takes a list of Char values, returning True if at least one of
them is '0' and False otherwise.

The zip function

The function zip pairs up the elements of two lists.

```
zip :: [a] -> [b] -> [(a,b)]

zip [1,2,3] "abc" => [(1,'a'),(2,'b'),(3,'c')]
zip [1,2,3] "ab" => [(1,'a'),(2,'b')]
zip [1,2] "abc" =>  [(1,'a'),(2,'b')]
```

If the lists are not of the same length, then the longer one's extra elements
do not appear in the result.

The zipWith function

The function zipWith is a function like map except that it takes two lists. Like zip, the longer list's extra elements are ignored.

```
zipWith :: (a -> b -> c) -> [a] -> [b] -> [c]

zipWith (+) [2,4..10] [1,3..10] => [3,7,11,15,19]
zipWith (*) [1,2,3] [1,2,3] => [1,4,9]
```

Exercise 8. Write a function that takes two lists of type [Maybe Int] and examines the pair of list heads before looking at the rest of the lists. It returns a list in which the Ints of each pair have been added if both are of the form Just n, preserving any Just n value otherwise. For example,

```
addJust [Just 2, Nothing, Just 3]
        [Nothing, Nothing, Just 5]
     => [Just 2, Nothing, Just 8]
```

The foldr function

Finally, there are two very powerful functions on lists that require some experience before you are really comfortable with them. Both make use of an *accumulator*, a value that becomes more and more like the final result of the computation.

Let's look at an accumulator more closely. Suppose that we apply (+) to the elements of the list [1,2,3] in this way:

```
(+) 1
    ((+) 2
        ((+) 3 0))
```

On the third line, there is an expression which forms the second argument of the function application that appears on the second line. That application in turn forms the second argument of the application appearing on the first line. In each case, the second argument of the (+) function is the result of a previous computation, handed on to the next application. This second argument is an accumulator.

The order in which the list elements are used is from right to left. First the 3 is added to 0, then the result of that addition is added to 2, the second element of the list. Finally 1, the first element of the list, is added to the result of the previous addition.

The foldr function does something very similar. It takes a function of two arguments, and the second argument is an accumulator. Initially, the accumulator has to have a value, so foldr receives an initial value as well. Its last argument is a list of values, and it returns the accumulated value. If we were to use foldr to implement the example above, we would write:

```
foldr (+) 0 [1,2,3]
```

The type of `foldr` is complex. Its function argument takes two values and returns the accumulator, so the type of this argument is `a -> b -> b`. The first value of the accumulator is the initial value, so its type is `b`. The type of the list argument is `[a]`, and the result type is `b`, the final accumulator value.

```
foldr :: (a -> b -> b) -> b -> [a] -> b
```

The accumulator makes it possible to do a surprising variety of things with `foldr`. For example, the `sum` function sums the numbers in its list argument. We can implement `sum` with `foldr` like this:

```
foldr (+) 0 [1,2,3,4,5]
  => (+) 1 ((+) 2 ((+) 3 ((+) 4 ((+) 5 0))))
  => 15
```

We can also implement (++) with `foldr`:

```
foldr (:) [3,4,5] [1,2]
  => (:) 1 ((:) 2 [3,4,5])
  => [1,2,3,4,5]
```

Other functions such as `and`, which returns True if its list argument contains only True values, and `or` (which is similar) can also be implemented using `foldr`:

```
foldr (&&) True [True, False, True] => False
```

```
foldr (||) False [True, False, True] => True
```

The . (composition) operator

Now let's take a look for just a moment at some more notation. The *composition* operator (`.`) allows us to create a pipeline of function applications, each of which is waiting for an argument. For example,

```
(toUpper.toLower) 'A' => 'A'
```

```
(toLower.toUpper) 'Z' => 'z'
```

In each case, the first (rightmost) function receives its argument, then gives its result to the function on the left. That function returns the final result of the application.

This notation also allows us to form function applications like this:

```
((:).toUpper) 'a' "bc" => "Abc"
```

Again, the first function receives the 'a' and returns 'A'. Then the *cons* function receives the 'A' and the string, and creates a new string.

With this in mind, new possibilities for using `foldr` become available. We can write `map`, or `elem`, a function that returns True if its first argument is in its list argument:

```
map toUpper "abc"
  = foldr ((:).toUpper) [] "abc"
     => "ABC"

elem 3 [1,2,3,4,5]
  = foldr ((||).(== 3)) False [1,2,3,4,5]
     => True
```

The `foldl` function

The `foldl` function also uses an accumulator, but it processes the list elements from left to right. We look again at applying `(+)` to the elements of the list `[1,2,3]`:

$$
\begin{array}{ccc}
(+) & & 3 \\
 & ((+) & 2) \\
 & ((+)\ 0\ 1) &
\end{array}
$$

As you can see, the first argument is the accumulator, and the *first* element of the list, not the last, is given to the accumulator initially.

In many cases, this makes no difference, and `foldl` returns the same result as `foldr`. For example, the following two expressions produce the same result because the multiplication operation is associative, and 1 is an identity for multiplication:

```
foldl (*) 1 [1,2,-3]
foldr (*) 1 [1,2,-3]
```

However, these expressions do not give the same result because subtraction is not associative:

```
foldl (-) 0 [1,2,3,4]

foldr (-) 0 [1,2,3,4]
```

It can be challenging to write applications of `foldr` and `foldl`. The best way to do this effectively is to do an example by hand until you get used to writing applications that work. This process is called 'expanding' an application. Here are some examples of expanding applications of `foldr` and `foldl`:

```
foldr (++) [] [[1],[2],[3]]
  => (++) [1] ((++) [2] ((++) [3] []))
  => (++) [1] ((++) [2] [3])
  => (++) [1] [2,3]
  => [1,2,3]

foldr (&&) False [True,False]
  => (&&) True ((&&) False False)
  => (&&) True False
  => False

foldl (-) 0 [1,2,3]
  => (-) ((-) ((-) 0 1) 2) 3
  => (-) ((-) -1 2) 3
  => (-) -3 3
  => -6

foldl max 0 [1,2,3]
  => max (max (max 0 1) 2) 3
  => max (max 1 2) 3
  => max 2 3
  => 3
```

Exercise 9. Expand the following applications:

```
foldr (** 2) 0 [1,2,3]

foldr seen False [3,2,1,4]

foldr spaces [] ["this" "is" "a" "sentence"]
```

1.9   Data Type Definitions

Tuples and lists are very useful, but there comes a point when you would like to define the shape of your data so that it fits the problem being solved. Haskell is particularly good at doing this because it supports *algebraic data types*, a flexible form of user-defined data structure. Furthermore, pattern matching, which we have already used for tuples and lists, can be used with the algebraic data types. Together, these allow you to define and use made-to-order structures.

Suppose that you wanted to specify an enumerated type in Haskell. For example, you would like a type that enumerates colours:

```
data Colour = Red | Orange | Yellow
            | Green | Blue | Violet
```

This declaration states that the `Colour` type contains each of the values `Red`, `Orange`, and so on. If these appeared in a list, as in [`Red, Yellow, Green`], the type of the list would be [`Colour`]. Each of these values is a *constructor*, and constructors are the only Haskell values that start with an upper case letter.

The `Bool` type which we have been using throughout is not actually a built-in type; it is defined as follows in the standard library:

```
data Bool = False | True
```

Now suppose that you would like values that contain fields, because information of some kind must be associated with each of the values. We might define an `Animal` type in which each value indicates an animal species and some additional information about a particular animal, such as its breed. The type would contain entries for dogs and cats, and a string for each of these which could be used to store their breed, but our entry for rats would not need a string for the breed because we are not particularly interested in rat breeding.

```
data Animal = Cat String | Dog String | Rat
```

Here are some values with type `Animals`:

```
Cat "Siamese"
Cat "Tabby"
Dog "Spaniel"
Dog "big and hungry"
Rat
```

As you can see, there is a field next to the `Cat` constructor, which can be accessed by a function defined on the type, using pattern matching.

Sometimes a more flexible approach is needed. Instead of storing a string with each cat or dog, we might want to store some arbitrary information. This can be accommodated by letting `Animal` take two *type variables*, a and b, which can stand for any data type:

```
data Animal a b
  = Cat a | Dog b | Rat
```

Type variables must start with lowercase letters. It is a common convention, but not required, to use a, b and so on for type variables. Type variables allow `Animal` to be used with arbitrary annotations, for example:

```
data BreedOfCat = Siamese | Persian | Moggie

Cat Siamese   :: Animal BreedOfCat b
Cat Persian   :: Animal BreedOfCat b
Cat "moggie"  :: Animal String b
Dog 15        :: Animal a Integer
Rat           :: Animal a b
```

Data types with type variables can be very useful. Suppose that you are writing a function that may succeed in computing its result, which has type a, but may also fail to get a result. A common example is a search function, which is looking for a value in a table, and which might fail. Instead of letting the program crash, it is better to return a value that says either 'I succeeded, and the result is *x*' or else 'I failed to get a result'. The Maybe type is just what we need:

```
data Maybe a = Nothing | Just a
```

Suppose we are writing a function to look up someone's telephone number, using a database represented as a list of pairs. The first component of each pair is a person's name, and the second component is their telephone number (an integer). It isn't a good idea to have this function simply return an integer; a search might be made for a person not in the database. The solution is to use Maybe:

```
phone_lookup :: [(String,Integer)] -> String -> Maybe Integer
```

Now the lookup function can be defined so that it always returns something sensible.

You may have tried some examples by this time, and noticed that Haskell will not print values of your new type. This is because it uses a function called show to convert a data value to a string that can be printed, and you have not constructed a version of show that it can use.

Fortunately, you do not have to define show yourself. Haskell will do it automatically for you when you end a data type definition with the words deriving Show, as in:

```
data Colour = Red | Orange | Yellow
            | Green | Blue | Violet
            deriving Show
```

Pattern matching can take place over values of any types, including algebraic data types defined by the user. For example, let's define a function that takes the result of a phone number search and produces a comprehensible string to be printed.

```
phone_message :: Maybe Integer -> String
phone_message Nothing = "Telephone number not found"
phone_message (Just x) = "The number is " ++ show x
```

Exercise 10. Define a data type that represents six different metals, and automatically creates versions of (==) and show.

Exercise 11. Suppose that you have some coins that have been sorted into piles, each of which contains only one kind of coin. Define a data type that can be used to represent the piles of coins.

Exercise 12. A *universal* type is one in which any type can be represented. Each different type is identified by its own constructor, which serves as a distinguishing tag. For example, here is a universal type that represents three different types of number:

```
data Number = INT Int | INTEGER Integer | FLOAT Float
              deriving (Eq, Show)
```

Define a universal type that contains booleans, characters and integers (Ints).

Exercise 13. Define a type that contains tuples of up to four elements.

Exercise 14. The quadratic equation $a \cdot x^2 + b \cdot x + c = 0$ has two roots, given by the formula

$$x = \frac{-b \pm \sqrt{b^2 - 4 \cdot a \cdot c}}{2 \cdot a},$$

as long as the discriminant (the expression under the square root sign) is non-negative. If the discriminant is negative the roots are complex. Define a function that finds the real solutions of the quadratic equation, and reports failure if they don't exist.

1.10 Type Classes and Overloading

There are some operations which can be used on several different types, but not on all types. For example, the (+) function can be applied to integers or floating point numbers (as well as several other kinds of number). This means that there are two completely different implementations of this function, one for integers and another for floating point. And that raises a question: what is the type of (+)? It would be too restrictive to specify

```
(+) :: Integer -> Integer -> Integer
```

because then it would be illegal to write 3.14+2.7. On the other hand, it would be too general to specify

```
(+) :: a -> a -> a
```

because this says that *any* type can be added, allowing nonsensical expressions like True+False and "cat"+"mouse". What we need to say is that (+) can be used to add two values of any type, provided that type is numeric. The (+) function actually has the following type:

```
(+) :: Num a => a -> a -> a
```

Num is the name of a *type class*, a set of types sharing a property. The members of the set Num include Int, Integer, Float, Double, and many others, and their essential property is that arithmetic makes sense on these types. However, types like Bool and String, where addition is meaningless, are *not* members of the class Num. In the type of (+), the notation Num a => is called a *class constraint* or a *context*, and it means that (+) can only be applied to arguments with types that belong to the Num set.

Haskell allows you to define new type classes, and to specify that a type is a member of a class. This is a powerful corner of the language, but we will not go into it in detail; see the references in Section 1.11 for a complete explanation.

To read this book, you don't need to be able to define new type classes or instances, but error messages sometimes mention type class constraints, so it is helpful to know what they mean.

There are a few commonly used type classes that are ubiquitous in Haskell; the most important are Num, Show and Eq. Show is the class of types that can be converted in a meaningful way into a character string. Most ordinary data values are in Show, but functions are not. Eq is the class of types that can be compared for equality.

When you are faced with a type error in a program, it is good to realise that values (with their class constraints) can migrate from a long way away. For example, if we have the definition

```
fun a b c = if a then b == c else False
```

then the type of fun has to reflect the fact that b and c have to be in the Eq class. This means that the fun function also has a type constrained by the Eq class, so the definition should be written as:

```
fun :: Eq b => a -> b -> b -> Bool
fun a b c = if a then b == c else False
```

If you forget to include the class constraint Eq b => in the type signature, the Haskell compiler will give an error message. The context on the type of fun declares that whatever calls fun and supplies it with arguments must ensure that there is a meaningful way to compare them.

You will inevitably come up against functions that have contexts in their types. The common sense rule is: if your function definition uses an overloaded operator (one with a type that has a context), then *its* type must contain that context as well. If your function has more than one such operator and the operator types have different contexts, then each new context must appear in the type of the function.

For example, suppose you want to write a function that checks whether a value appears in a list, and returns a corresponding error message. Here's a definition:

```
detect :: (Eq a, Show a) => a -> [a] -> String
```

```
detect v list =
  if elem v list
    then "List contains " ++ show a
    else "List does not contain" ++ show a
```

The function `elem` has a type that contains the context `Eq a` because it uses the overloaded operator `==`. The type of `detect` now has to have this context, and it also needs the `Show a` context, because it uses the overloaded operator `show`.

1.11 Suggestions for Further Reading

A good source of information on the web about Haskell, and functional programming in general, is the Haskell Home Page: `www.haskell.org`. This contains pointers to a variety of relevant books and papers, as well as the language reference manuals.

Several good textbooks on Haskell are available: *Introduction to Functional Programming using Haskell*, by Bird [3]; *An Introduction to Functional Programming Systems Using Haskell*, by Davie [7]; and *Haskell: The Craft of Functional Programming*, by Thompson [30].

The book by Hudak, *The Haskell School of Expression: Learning Functional Programming through Multimedia* [17], shows how to use Haskell through a series of applications to graphics, animations and music.

The use of equations rather than assignments gives functional programming a very different style from imperative programming. *Purely Functional Data Structures* [22], by Okasaki, explores this topic in depth, and is an excellent intermediate level text on functional programming.

1.12 Review Exercises

Exercise 15. Define a function

```
showMaybe :: Show a => Maybe a -> String
```

that takes a `Maybe` value and prints it.

Exercise 16. A Bit is an integer that is either 0 or 1. A Word is a list of bits that represents a binary number. Here are some binary values that can be represented by Words:

```
[1,0] => 2
[1,0,0,1] => 9
[1,1,1] => 7
```

We can define functions that are the Bit equivalent of `or` and `and` as follows:

```
bitOr :: Int -> Int -> Int
bitOr 0 0 = 0
bitOr x y = 1

bitAnd :: Int -> Int -> Int
bitAnd 1 1 = 1
bitAnd x y = 0
```

Now it is possible to take the 'bitwise' *and* of two words as follows:

```
bitwiseAnd [1,0,0] [1,0,1]
 => [bitAnd 1 1, bitAnd 0 0, bitAnd 0 1]
 => [1,0,0]

bitwiseAnd [0,0,0] [1,1,0]
 => [0,0,0]
```

Write a function `bitwiseAnd` that takes two Words and creates a third Word that is the bitwise *and* of the two Words.

Exercise 17. Suppose that you are to define a type that contains lists of any type, but only up to a depth of three. The proposed solution is:

```
data Lists1 a = L1 [a]
                  deriving (Eq, Show)

data Lists2 a = L2 [Lists1 a]
                  deriving (Eq, Show)

data Lists3 a = L3 [Lists2 a]
                  deriving (Eq, Show)
```

Is it possible to represent a list of depth greater than three with these definitions?

Exercise 18. Each of the following expressions has a type error. Change the expression so that the type error no longer occurs.

```
[1, False]              '2' ++ 'a'
[(3,True), (False,9)]   2 == False
'a' > "b"               [[1],[2],[[3]]]
```

Exercise 19. What caused the type error in this definition and application?

```
f :: Num a => (a,a) -> a
f (x,y) = x + y

f (True,4)
```

Exercise 20. Why does this definition produce an error when used?

```
f :: Maybe a -> [a]
f Nothing = []
```

```
f (Just 3)
```

Exercise 21. Write a list comprehension that takes a list of Maybe values and returns a list of the Just constructor arguments. For example,

```
[Just 3, Nothing, Just 4] => [3,4]
```

Exercise 22. Using a list comprehension, write a function that takes a list of Int values and an Int value n and returns those elements in the list that are greater than n.

Exercise 23. Write a function

```
f :: [Int] -> Int -> [Int]
```

that takes a list of Int values and an Int and returns a list of indexes at which that Int appears.

Exercise 24. Write a list comprehension that produces a list giving all of the positive integers that are *not* squares in the range 1 to 20.

Exercise 25. Write a function that uses `foldr` to count the number of times a letter occurs in a string.

Exercise 26. Write a function using `foldr` that takes a list and removes each instance of a given letter.

Exercise 27. Using `foldl`, write a function

```
rev :: [a] -> [a]
```

that reverses its list argument.

Exercise 28. Using `foldl`, write a function

```
maybeLast :: [a] -> Maybe a
```

that takes a list and returns the last element in it if there is one, otherwise it returns Nothing.

Chapter 2

Propositional Logic

Logic provides a powerful tool for reasoning correctly about mathematics, algorithms and computers. It is used extensively throughout computer science, and you need to understand its basic concepts in order to study many of the more advanced subjects in computing. Here are just a few examples, spanning the entire range of computing applications, from practical commercial software to esoteric theory:

- In *software engineering* it is good practice to specify what a system should do before starting to code it. Logic is frequently used for software specifications.

- In *safety-critical applications*, it is essential to establish that a program is correct. Conventional debugging isn't enough—what we want is a proof of correctness. Formal logic is the foundation of program correctness proofs.

- In *information retrieval*, including Web search engines, logical propositions are used to specify the properties that should (or should not) be present in a piece of information in order for it to be considered relevant.

- In *artificial intelligence*, formal logic is sometimes used to simulate intelligent thought processes. People don't do their ordinary reasoning using mathematical logic, but logic is a convenient tool for implementing certain forms of reasoning.

- In *digital circuit design and computer architecture*, logic is the language used to describe the signal values that are produced by components. A common problem is that a first-draft circuit design written by an engineer is too slow, so it has to be transformed into an equivalent circuit which is more efficient. This process is often quite tricky, and logic provides the framework for doing it.

35

- In *database systems*, complex queries are built up from simpler components. It's essential to have a clear, precise way to express such queries, so that users of the database can be sure they're getting the right information. Logic is the key to expressing such queries.

- In *compiler construction*, the typechecking phase must determine whether the program being translated uses any variables or functions inconsistently. It turns out that the method for doing this is similar to the method for performing logical inference. As a result, algorithms that were designed originally to perform calculations in mathematical logic are now embedded in modern compilers.

- In *programming language design*, one of the most commonly used methods for specifying the meaning of a computer program is the *lambda calculus*, which is actually a formal logical system originally invented by a mathematician for purely theoretical research.

- In *computability theory*, logic is used both to specify abstract machine models and to reason about their capabilities. There has been an extremely close interaction between mathematical logicians and theoretical computer scientists in developing a variety of machine models.

In this chapter, we discuss the difficulties with informal logical reasoning in English, and we show how to avoid those difficulties with formal logic. There are several different kinds of formal logic, and for now we will consider just the simplest one, called *propositional logic*. After looking at the language of propositional logic, we will consider in detail three completely different mathematical systems for reasoning formally about propositions: truth tables, natural deduction and Boolean algebra.

Truth tables define the meanings of the logical operators, and they can be used to calculate the values of expressions and prove that two propositions are logically equivalent. Since truth tables directly express the underlying meaning of propositions, they are a *semantic* technique. Truth tables are easy to understand for small problems, but they become impossibly large for most realistic problems.

Natural deduction is a formalisation of the basic principles of reasoning. It provides a set of *inference rules* that specify exactly what new facts you are allowed to deduce from some given facts. There is no notion of the 'value' of propositions; everything in this system is encapsulated in the inference rules. Since the rules are based on the forms of propositions, and we don't work with truth values, inference is a purely *syntactic* approach to logic. Many recently developed techniques in programming language theory are based on more advanced logical systems that are related to natural deduction.

Boolean algebra is another syntactic formalisation of logic, using a set of equations—the laws of Boolean algebra—to specify that certain propositions are equal to each other. Boolean algebra is an axiomatic approach, similar to elementary algebra and geometry. It provides an effective set of laws for manipulating propositions, and is an essential tool for digital circuit design.

2.1 The Need for Formalism

Formal logic was first developed by the ancient Greeks, who wanted to be able to reason carefully about statements in natural language. They were fascinated by the idea of statements which are known to be true with absolutely no doubt whatsoever. However, they quickly realised that logical reasoning is difficult and unreliable when using a natural language like Greek (or English!). All sorts of ambiguities arise and it's hard to keep sight of the main line of reasoning without getting confused. We will begin by looking at some of these difficulties, and how to get around them using *propositional variables*.

Suppose a friend says 'The sun is shining and I feel happy'. At first sight the meaning is obvious, but when you think about the sentence more carefully it isn't so clear. Perhaps your friend likes sunny days and feels happy *because* the sun is shining. It sounds like there is some connection between the two parts of the sentence, so in this context *and* means *and therefore*. But this reasoning depends on our experience with bright weather and happiness, and it has nothing to do with the logical structure of the sentence. Now consider another example: 'Cats are furry and elephants are heavy'. This has exactly the same structure as the preceding example, but nobody would assume that elephants are heavy *because of* the furriness of cats. In this case, *and* means *and also*. The word *and* has several subtly-different meanings, and we choose the appropriate meaning using our knowledge of the world. Unfortunately, this means we can't even rely on a simple word like *and* while reasoning in English.

Furthermore, there are many other problems in working out the precise meanings of English sentences. For example, 'The sun is shining' is true some days and false other days. The meaning of 'That cloud looks like a motor bike' depends on who says it, and which cloud they are pointing at. The list of such problems seems to be endless, and you can read about them in books on linguistics and philosophy.

There is no way to solve all the ambiguities of English. Who would want to do that anyway? The subtle nuances in natural language are not necessarily bad: they lead to much of the richness and expressiveness of literature. Yet they certainly can get in the way of logical thinking.

Instead of attempting the impossible—totally reliable reasoning in natural language—we need to *separate* the logical structure of an argument from all the connotations of the English. We do this using *propositions*.

A proposition is just a symbolic variable whose value must be either True or False, and which stands for an English statement which could be either true or false. The crucial point here is that a proposition *must* be either True or False; there is no room for shades of meaning or interpretation. Usually we'll use A, B, C, D etc. as propositional variables, but any variable name would

do. For example, we can define some propositional variables to stand for the
following English statements:

$$
\begin{aligned}
A &= \text{The sun is shining.} \\
B &= \text{I feel happy.} \\
C &= \text{Cats are furry.} \\
D &= \text{Elephants are heavy.}
\end{aligned}
$$

The next step is to translate a complete English sentence into a mathematical statement which contains nothing but propositional variables (A, B etc.) and logical operators (*and, or, not, implies*).

| The sun is shining and I feel happy. | $\Longrightarrow$ | A and B |
| Cats are furry and elephants are heavy. | $\Longrightarrow$ | C and D |

The translation step is absolutely crucial, because it removes all the ambiguities of English and all our knowledge and experience of the world (like sunlight bringing happiness), and it leaves us with nothing but propositional variables and logical operators.

Sometimes it isn't clear how to translate an English statement into propositional logic. What about the sentence, 'It is raining but Jim is happy'—does this mean the same thing as 'It is raining *and* Jim is happy', or does the use of *but* indicate a different meaning? Such questions fall outside the realm of mathematics, and you just have to figure out what the English means, or ask for a clarification. At least there is only one time when we need to worry about the subtleties of natural language—during the translation process—and we can forget about it thereafter.

A proposition must be either true or false; it cannot be 'maybe' or 'sometimes' or 'yes, but...'. If you translate an opinion like 'Cats are better than dogs because they purr' into a propositional variable P, then within the mathematics you'll just have to accept P as being true or false, and you won't be able to get at anything *inside* P, such as the reason that cats are better than dogs (indeed, you can't even get inside P to find out that it is about cats and dogs, not about turtles and rabbits).

There are many statements that cannot be represented by propositions because they require context in order to make their meaning clear. For example, if we define A to represent 'That cloud looks like a motor bike', then it must be clearly established in the English *which* cloud looks like a motor bike, and *who* thinks that. There are more complex logical systems that incorporate time, but propositional logic doesn't do that.

2.2 The Basic Logical Operators

Logical operators correspond to English words like *and, or, not* and *therefore*, providing a way to build complex propositions from simpler ones. This section defines the exact meaning of the logical operators, and shows how to translate between English statements and mathematical propositions.

2.2.1 Logical And ($\wedge$)

The *logical and* operator corresponds to the English conjunction 'and': it is used to claim that two statements are both true. Sometimes it is called 'logical conjunction', and its mathematical symbol is $\wedge$.

The proposition $A \wedge B$ is simply a statement that A is True and also B is True. It doesn't have any subtle connotations; in particular, it doesn't mean that there is any connection between A and B.

It's vitally important to remember that people can say things that are untrue! If someone tells you '$A \wedge B$', you have to bear in mind two possibilities:

- Their statement was correct.

- They lied to you.

If their statement was true, then both A and B are indeed true. However, if they lied to you[1] then you don't actually know about the truth of A or B.

Since we can't simply accept every statement as being true, we need a way to calculate whether a statement is true based on its constituent parts. For example if you already know that A is true and B is true, then the statement 'A and B' is certainly true. But if you know that A is false (or that B is false), then you know that the statement '$A \wedge B$' is false.

The mathematical symbol for logical *and* is $\wedge$. This symbol is shorter than '*and*', and it is clearly a mathematical operator—there is no danger of confusing $\wedge$ with the various vague meanings of the English word 'and'.[2]

Think of *and* as an operator over logical propositions, just as $+$ is an operator over numbers. We can define the meaning of *and* by considering whether '$A \wedge B$' is true for all possible values of A and all possible values of B. Such a listing is called a truth table, and here is the definition of logical *and*:

A	B	$A \wedge B$
False	False	False
False	True	False
True	False	False
True	True	True

---

[1] A Los Angeles car salesman famous for his flamboyant television advertisements once ran a commercial claiming, 'We lose money on every car we sell, but we make it up on volume.'

[2] In the elderly (but still popular!) programming language Cobol, you do arithmetic with statements like Add a to b giving x instead of the more usual x := a+b. The designers of Cobol believed that $+$ is mathematical, and therefore difficult, while 'Add ... giving ...' is English, and therefore easy. The Cobol notation is at least partly readable by nonprogrammers, which may be valuable in commercial applications, but it becomes unwieldy for large scale calculations. Once you get used to it, mathematical notation is *easier* than English!

Each time False is an argument value, ∧ returns False. Only when both arguments are True does it return True.

Various notations are used for the two truth values. Here we have used the full names True and False, but those become tedious to write when there are a lot of entries in the table. Another common notation is to use T and F. The usual notation in digital circuit design is to use 1 for True and 0 for False. Apart from being concise, this has the great advantage that 0 and 1 are more easily distinguishable than T and F, making the truth tables more readable. You may use any of these notations. However, if you use 0 and 1 as your notation for the truth values, be sure to remember that these are not numbers. For the sake of comparison, here is how the preceding truth table looks like with the 0, 1 notation.

A	B	$A \wedge B$
0	0	0
0	1	0
1	0	0
1	1	1

Given any two propositions, you can build a bigger one by connecting them with ∧. For example, from the propositional variables A, B and C we can build an endless list of more complex propositions, including the following:

$$A \wedge B$$
$$A \wedge (B \wedge B)$$
$$(A \wedge B) \wedge C$$
$$(A \wedge B) \wedge (B \wedge C)$$

2.2.2 Inclusive Logical Or (∨)

The *logical or* operator corresponds to the most common usage of the English word *or*. It takes two arguments and returns True if either argument (or both) is True; otherwise it returns False. Other commonly used names for this operation are *logical disjunction* and *inclusive or*, and the symbol for the operation is ∨. Here is the truth table defining ∨:

P	Q	$P \vee Q$
False	False	False
False	True	True
True	False	True
True	True	True

The English word 'or' has several different meanings. The *inclusive or* function corresponds to the simplest of these: if $A \vee B$ is true, then perhaps A is true, perhaps B is true, perhaps both are true, but you know they can't both be false. However, this is *all* that $A \vee B$ means. It does not indicate any connection between A and B.

2.2.3 Exclusive Logical Or (⊗)

In English, we often think of '*A* or *B*' as meaning 'either *A* or *B*—one or the other, but not both', excluding the possibility that *A* is true at the same time as *B*. A typical example is 'It will be bright and sunny at the picnic, or I will go home.' This meaning of the word 'or' is called *exclusive or* because if one of the alternatives is true, this excludes the possibility that the other is also true. Often the 'either ... or' construction indicates that the exclusive or is intended; for example: 'Either you pay me, or I will sue you.' However, you can't always rely on the presence of 'either'. The symbol for exclusive or is ⊗. Its truth table differs from the ∨ truth table only in the last line.

P	Q	$P \otimes Q$
False	False	False
False	True	True
True	False	True
True	True	False

Since the English word 'or' sometimes means inclusive or and sometimes exclusive or, you have to be particularly careful when translating English sentences containing 'or' into propositions.

In many applications of mathematical logic, inclusive or plays a central role while exclusive or isn't too important. In many books and papers, the word 'or' is taken to mean inclusive or, and the exclusive variety isn't mentioned at all. When logic is applied to digital circuit design, both kinds of or are used heavily.

2.2.4 Logical Not (¬)

The English word 'not' is used to assert that a statement is false. It corresponds to the *logical not* (also called *logical negation*) operator, whose symbol is ¬. It takes one argument and returns the opposite truth value:

A	$\neg A$
False	True
True	False

2.2.5 Logical Implication (→)

The logical implication operator → corresponds to conditional phrases in English. For example, an English sentence in the form 'If *A* is true, then *B* is true.' would be translated into the proposition $A \rightarrow B$. Logical implication is also closely related to the **if ... then ... then** construct in programming languages. As usual, the precise definition is given by a truth table:

A	B	$A \rightarrow B$
False	False	True
False	True	True
True	False	False
True	True	True

Suppose you know that a proposition $A \rightarrow B$ is true. This says that *if* A is true, then B *must also* be true. But that's all it means—in particular, it doesn't tell you whether A is actually true. A common pitfall in informal English debate is to jump to the conclusion that A is true, when the speaker has merely said 'if A is true, then B is also true.' In the following example, $A \rightarrow B$ does *not* tell you what the weather will be like today!

A	It is sunny today.
B	There will be a picnic.
$A \rightarrow B$	If it is sunny today, then there will be a picnic.

There is a subtle but crucial difference between $\rightarrow$ and the corresponding English sentences. The last English sentence in the example above suggests a cause-and-effect relationship between the weather and the picnic: the picnic might be cancelled *because of* the rain. In contrast, $A \rightarrow B$ says nothing at all about any connection between A and B; it means nothing more than what the truth table says. Consider the following example:

A	The moon orbits the earth.
B	The sun is hot.
$A \rightarrow B$	If the moon orbits the earth, then the sun is hot.

These statements are all true! Logically, this is identical to the previous example. Since A and B are both true, the definition of $\rightarrow$ says that $A \rightarrow B$ is also true. However, the English translation sounds strange: it is true that the moon orbits the earth, and the sun is surely hot, but the *reason* the sun is hot has nothing to do with the moon. The English sentence is misleading, and you wouldn't want to use it in a debate, but there isn't anything wrong with the logical proposition. *Logical implication says nothing about cause-and-effect relationships.*

Another point about implication sometimes causes confusion: many people find the first line of the implication truth table surprising. If A is false, then why should $A \rightarrow B$ be true, when B is actually untrue? Let's see how this might translate into English, picking an example where A and B are both false:

A	The sun is cold.
B	The moon is made of green cheese.
$A \rightarrow B$	If the sun is cold, then the moon is made of green cheese.

Here, A and B are both false, but the definition says that $A \rightarrow B$ is true. Yet why should the English translation of $A \rightarrow B$ be true?

Even though it may seem strange to you to define False $\rightarrow$ False to be True , you probably already understand the idea underlying the definition: it reflects a colourful way to express skepticism. Suppose, for example, that your friend claims 'Fifty people came to my party last night', but you're sure there were only twenty. You might retort, 'If fifty people went to your party, then I'm the king of China'. Your reply has the form $A \rightarrow$ False, where A means fifty people went to the party, since you aren't the king of China. Furthermore, *the statement you are making is true*. Thus you are asserting $A \rightarrow$ False = True, and a quick study of the truth table for $\rightarrow$ shows that A must be False.

English sentences contain all sorts of connotations but logical propositions mean nothing more than what the defining truth tables say. The truth table for $\rightarrow$ is just a definition. It is meaningless to argue about whether the definition given above is 'correct'; definitions aren't either right or wrong, they are just definitions.

The pertinent question to ask is *why is it convenient to define $\rightarrow$ so that* False $\rightarrow$ False *is true?* Perhaps the most honest answer is just to say that logicians have been refining their definitions for hundreds of years, and based on their experience, they feel this is the best way to define $\rightarrow$. However, here's a more computer-oriented way to think about it: consider the programming language statement **if** A **then** B. If A turns out to be true, then the statement B will be executed, but if A is false, then the machine won't even look at B. It makes no difference at all what B says. The definition of $\rightarrow$ captures this indifference to B in the cases where A is false.

2.2.6   Logical Equivalence ($\leftrightarrow$)

The logical equivalence operator $\leftrightarrow$ is used to claim that two propositions have the same value—either both are true, or both are false. The proposition $A \leftrightarrow B$ might be translated into an English sentence like 'saying A is just the same as saying B.' The definition of $\leftrightarrow$ is

A	B	$A \leftrightarrow B$
False	False	True
False	True	False
True	False	False
True	True	True

The statement $A \leftrightarrow B$ could also be expressed by writing $(A \rightarrow B) \wedge (B \rightarrow A)$. Sometimes $\leftrightarrow$ is simply considered to be an abbreviation for conjunction of two ordinary implications, rather than a fundamental logical operator.

There are two distinct mathematical ways to claim that the propositions A and B have the same value. One way is to write the equation $A = B$, and the other way is to claim that the proposition $A \leftrightarrow B$ has the value True. Logical equivalence is similar to *but not the same as* ordinary equality. The difference between them is important, and we'll say more about it later.

2.3   The Language of Propositional Logic

Before spending a lot of effort arguing about a statement somebody claims to be true, it's prudent to make sure the statement actually makes sense, and to check that all the participants in the debate agree about what it means. It might be fun to discuss 'International trade creates jobs', but there's no point wasting time on 'Quickly raspberries running green'. That one isn't either true or false—it's just gibberish.

A big advantage of using logic rather than English is that we can define the language of propositional logic formally, precisely and unambiguously. This allows us to check that a proposition makes sense before expending any further work deciding whether it's true. Since there is no formal definition of natural languages, it's fundamentally impossible to do that for English. It isn't even possible in all cases to decide definitively whether a statement in a natural language is grammatical: English has no formal grammar.

A proposition which 'makes sense' is called a *well-formed formula*, which is often abbreviated WFF (pronounced 'woof'). Every WFF has a well-defined meaning, and you can work out whether it's true or false given the values of the propositional variables. For example, $(A \rightarrow (B \wedge (\neg A)))$ is a well-formed formula, so we can calculate what it means, and it's worth spending a little time to decide whether there are any possible values of A and B for which it's true. (It's True if A is False.) On the other hand, what about $\vee AB \neg C$? This isn't a WFF. It's just nonsense, so it has no truth value.

Many programming languages (including Fortran, Pascal, Ada, Haskell and many others) provide a Boolean data type[3] for use with control statements such as **if** statements, **while** loops and so on. A variable of type Boolean has just two values, True and False, so it corresponds exactly to a propositional variable. These programming languages also provide some or all of the $\wedge$, $\vee$, $\neg$, $\rightarrow$ and $\leftrightarrow$ operators. Saying that a logical proposition is well-formed is just like saying that a Boolean expression in a programming language is syntactically valid.

There are also many programming languages (C, Lisp and others) which lack a Boolean type, using numbers instead. This is a loose approach to syntax, quite different from the precise rules for WFFs in logic. It is sometimes convenient, but it also leads to problematical definitions about whether expressions like $\sqrt{\sin x}$ are true or false, and it prevents the compiler from producing helpful type error messages if you accidentally use a non-Boolean expression where only a Boolean would make sense.

2.3.1   The Syntax of Well-Formed Formulas

A term in propositional logic is sometimes called a *formula*, and a term that is constructed correctly, following all the syntax rules, is called a *well-formed formula*, abbreviated *WFF*.

---

[3]Different languages use different names for the Boolean type; common choices include Bool, Boolean and Logical.

Soon we will study in detail how to reason about the meanings of WFFs. However, we will never consider the meaning of ill-formed formulas; these are simply considered to be meaningless. For example, the formula $(P \land (\neg P))$ is well-formed; this means that we can go on to consider its meaning (this WFF happens to be False). However, the formula $P\lor \rightarrow Q$ violates the syntax rules of the language, so we refuse to be drawn into debates about whether it is True or False. The situation is similar to that of programming languages: we can talk about the behaviour of a program that is syntactically correct (and the behaviour might be correct or incorrect), but we never talk about the behaviour of a program with syntax errors—the compiler would refuse to translate the program, so there is no run-time behaviour.

The set of well-formed formulas (WFFs) is defined by saying precisely how you can construct them.

- The constants False and True are WFFs.
  The only examples are: False,   True.

- Any propositional variable is a WFF.
  Examples: P, Q, R.

- If a is a WFF, then $(\neg a)$ is a WFF.
  Examples: $(\neg P)$, $(\neg Q)$, $(\neg(\neg P))$.

- If a and b are WFFs, then $(a \land b)$ is a WFF.
  Examples: $(P \land Q)$, $((\neg P) \land (P \land Q))$.

- If a and b are WFFs, then $(a \lor b)$ is a WFF.
  Examples: $(P \lor Q)$, $((P \land Q) \lor (\neg Q))$.

- If a and b are WFFs, then $(a \rightarrow b)$ is a WFF.
  Examples: $(P \rightarrow Q)$, $((P \land Q) \rightarrow (P \lor Q))$.

- If a and b are WFFs, then $(a \leftrightarrow b)$ is a WFF.
  Examples: $(P \leftrightarrow Q)$, $((P \land Q) \leftrightarrow (Q \land P))$.

- Any formula that cannot be constructed using these rules is *not* a WFF.
  Examples: $(P \lor \land Q)$, $P \rightarrow \neg$.

The rules may be used repeatedly to build nested formulas. This process is called *recursion*, and is the subject of Chapters 5 and 6. For example, here is a demonstration, using recursion, that $(P \rightarrow (Q \land R))$ is a WFF:

1. P, Q and R are propositional variables, so they are all WFFs.

2. Since Q and R are WFFs, $(Q \land R)$ is also a WFF.

3. Since P and $(Q \land R)$ are both WFFs, so is $(P \rightarrow (Q \land R))$.

2.3.2   Precedence of Logical Operators

Well formed formulas are fully parenthesised, so there is no ambiguity in their interpretation. Often, however, it's more convenient to omit some of the parentheses for the sake of readability. For example, we would prefer to write $P \to \neg Q \wedge R$ rather than $(P \to ((\neg Q) \wedge R))$.

The syntax rules given below define what an expression means when some of the parentheses are omitted. These conventions are analogous to those of elementary algebra, as well as most programming languages, where there is a precedence rule that says $a + b \times c$ means $a + (b \times c)$ rather than $(a + b) \times c$. The syntax rules for propositional logic are straightforward:

1. The most tightly binding operator is $\neg$. For example, $\neg P \wedge Q$ means $(\neg P) \wedge Q$. Furthermore, $\neg\neg P$ means $\neg(\neg P)$.

2. The second highest precedence is the $\wedge$ operator. In expressions combining $\wedge$ and $\vee$, the $\wedge$ operations come first. For example, $P \vee Q \wedge R$ means $P \vee (Q \wedge R)$. If there are several $\wedge$ operations in a sequence, they are performed left to right; for example, $P \wedge Q \wedge R \wedge S$ means $(((P \wedge Q) \wedge R) \wedge S)$. This property is described by saying '$\wedge$ associates to the left.'

3. The $\vee$ operator has the next level of precedence, and it associates to the left. For example, $P \wedge Q \vee R \vee U \wedge V$ means $(((P \wedge Q) \vee R) \vee (U \wedge V))$. This example would be more readable if a few of the parentheses are retained: $(P \wedge Q) \vee R \vee (U \wedge V)$.

4. The $\to$ operator has the next lower level of precedence. For example, $P \wedge Q \to P \vee Q$ means $(P \wedge Q) \to (P \vee Q)$. The $\to$ operator associates to the right; thus $P \to Q \to R \to S$ means $(P \to (Q \to (R \to S)))$.

5. The $\leftrightarrow$ operator has the lowest level of precedence, and it associates to the right, but we recommend that you use parentheses rather than relying on the associativity.

In general it's a good idea to omit parentheses that are clearly redundant, but to include parentheses that improve readability. Too much reliance on these syntax conventions leads to inscrutable expressions.

2.3.3   Object Language and Meta-Language

Propositional logic is a precisely defined language of well-formed formulas. We need another richer language to use when we are reasoning *about* well-formed formulas. This means we will be working simultaneously with two distinct languages which must be kept separate. The WFFs of propositional logic will be called the *object language* because the objects we will be talking about are sentences in propositional logic. The algebraic language of equations, substitutions and justifications is essentially just ordinary mathematics; we will call it the *metalanguage* because it 'surrounds' the object language and can be used to talk *about* propositions.

This distinction between object language and meta language is common in computer science. For example, there are many programming languages, and we need both to write programs *in* them and also to make statements *about* them.

We have already seen all the operators in the propositional logic object language; these are ¬, ∧, ∨, → and ↔. Later we will use several operators that belong to the meta language; these are ⊨ (which will be used in Section 2.4 on truth tables); ⊢ (which will be used in Section 2.5 on natural deduction); and = (which will be used in Section 2.7 on Boolean algebra). After studying each of those three major systems, we will look briefly at some of the meta logical properties of ⊨, ⊢ and = in Section 2.9 on metalogic.

2.3.4 Computing with Boolean Expressions

It's quicker and more reliable to compute large logical expressions with a computer than doing it by hand. In Haskell, logical variables have type `Bool`, and there are two constants of type `Bool`, namely `True` and `False`. Haskell contains built-in operators for a few of the logical operations, and the software tools provided with this book define the rest:

- ¬x is written in Haskell as `not x`;

- $a \wedge b$ is written either as `a && b` or as `a /\ b`;

- $a \vee b$ is written either as `a || b` or as `a \/ b`;

- $a \rightarrow b$ is written as `a ==> b`;

- $a \leftrightarrow b$ is written as `a <=> b`.

Exercise 1. Check your understanding of *or*, *and* and *not* by deciding what value each of these expressions has, and then evaluating it with the computer:

(a) `False /\ True`

(b) `True \/ (not True)`

(c) `not (False \/ True)`

(d) `(not (False /\ True)) \/ False`

(e) `(not True) ==> True`

(f) `True \/ False ==> True`

(g) `True ==> (True /\ False)`

(h) `False ==> False`

(i) (not False) <=> True

(j) True <=> (False ==> False)

(k) False <=> (True /\ (False ==> True))

(l) (not (True \/ False)) <=> False /\ True

2.4   Truth Tables: Semantic Reasoning

Truth tables provide an easy method for reasoning about propositions (as long as they don't contain too many variables). The truth table approach treats propositions as expressions to be evaluated. All the expressions have type Bool, the constants are True and False, and the variables $A, B, \ldots$ stand for unknowns that must be either True or False.

When you write out truth tables by hand, it's really too tedious to keep writing True and False. We recommend abbreviating True as 1 and False as 0; as pointed out earlier, T and F are also common abbreviations but are harder to read. However, it should be stressed that 0 and 1 are just abbreviations for the logical constants, whose type is Bool. Truth tables don't contain numbers. Only a few truth tables will appear below, so we will forgo the abbreviations, and stick safely with True and False. (However, when truth tables are used for reasoning about digital circuits, it is standard practice to use 0 and 1.)

2.4.1   Truth Table Calculations and Proofs

We have already been using truth tables to define the logical operators, so the format should be clear by now. Nothing could be simpler than using a truth table to evaluate a propositional expression: you just calculate the proposition for all possible values of the variables, write down the results in a table, and see what happens. Generally it's useful to give several of the intermediate results in the truth table, not just the final value of the entire proposition. For example, let's consider the proposition $((A \to B) \land \neg B) \to \neg A$ and find out when it's true. On the left side of the table we list all the possible values of A and B, on the right side we give the value of the full proposition, and in the middle (separated by double vertical lines) we give the values of the subexpressions.

A	B	$A \to B$	$\neg B$	$(A \to B)$ $\land \neg B$	$\neg A$	$((A \to B) \land \neg B)$ $\to \neg A$
False	False	True	True	True	True	True
False	True	True	False	False	True	True
True	False	False	True	False	False	True
True	True	True	False	False	False	True

Many propositions may be either True or False, depending on the values of their variables; all of the propositions between the double vertical lines in the table above are like that. Special names are given to propositions where this is not the case:

Definition 1. A *tautology* is a proposition that is always True, regardless of the values of its variables.

Definition 2. A *contradiction* is a proposition that is always False, regardless of the values of its variables.

You can find out whether a proposition is a tautology or a contradiction (or neither) by writing down its truth table. If a column contains nothing but True, the proposition is a tautology; if there is nothing but False it's a contradiction, and if there is a mixture of True and False the proposition is neither. For example, $P \vee \neg P$ is a tautology, but $P \wedge \neg P$ is a contradiction, and the following truth table proves it:

P	$\neg P$	$P \vee \neg P$	$P \wedge \neg P$
False	True	True	False
True	False	True	False

There is a special notation for expressing statements about propositions. Since it's used to make statements *about* propositions, this notation belongs to the meta-language, and it uses the meta operator $\models$, which is often pronounced 'double-turnstile' (to distinguish it from $\vdash$, which you'll see later, and which is pronounced 'turnstile').

Definition 3. The notation $P_1, P_2, \ldots, P_n \models Q$ means that if all of the propositions $P_1, P_2, \ldots, P_n$ are true then the proposition Q is also true.

The $\models$ meta-operator makes a statement about the actual meanings of the propositions; it's concerned with which propositions are True and which are False. Truth tables can be used to prove statements containing $\models$. The meaning of a proposition is called its *semantics*. The set of basic truth values (True and False), along with a method for calculating the meaning of any well-formed formula, is called a *model of the logical system*.

There may be any number n of propositions $P_1, P_2, \ldots, P_n$ in the list of assumptions. If $n = 1$ then the statement looks like $P \models Q$. A particularly important case is when $n = 0$, so there are no assumptions at all, and the statement becomes simply $\models Q$. This means that Q is always true, regardless of the values of the propositional variables inside it; in other words, Q is a tautology.

2.4.2 Limitations of Truth Tables

The truth table method is straightforward to use; it just requires some brute-force calculation. You don't need any insight into *why* the proposition you're working on is true (or false, or sometimes true). This is both a strength and a weakness of the method. It's comforting to know that you can reliably crank out a proof, given enough time. However, a big advantage of the other two logical systems we will study, natural deduction and Boolean algebra, is that they give much better insight into *why* a theorem is true.

The truth table method requires one line for each combination of variable values. If there are k variables, this means there are 2^k lines in the table. For one variable, you get $2^1 = 2$ lines (for example, the truth table for the $\neg$ operator). For two variables, there are $2^2 = 4$ lines (like the definitions of $\vee$, $\wedge$, $\rightarrow$ and $\leftrightarrow$). For three variables, there are $2^3 = 8$ lines which is about as much as anybody can handle.

Since the number of lines in the table grows exponentially in the number of variables, the truth table method becomes unwieldy for most interesting problems. In Chapter 10 we will prove a theorem whose truth table has 2^{129} lines. That number is larger than the number of atoms in the known universe, so you'll be relieved to know that we will skip the truth table, and use more powerful methods.

2.4.3   Computing Truth Tables

The software tools for this book contain functions that make it easy to compute truth tables automatically; see the online documentation.

Exercise 2. Use the truth table functions to determine which of the following formulas are tautologies.

(a) $(\text{True} \wedge P) \vee Q$

(b) $(P \vee Q) \rightarrow (P \wedge Q)$

(c) $(P \wedge Q) \rightarrow (P \vee Q)$

(d) $(P \vee Q) \rightarrow (Q \vee P)$

(e) $((P \vee Q) \wedge (P \vee R)) \leftrightarrow (P \wedge (Q \vee R))$

2.5   Natural Deduction: Inference Reasoning

Natural deduction is a formal logical system which allows you to reason directly with logical propositions using *inference*, without having to substitute truth values for variables or evaluate expressions. Natural deduction provides a solid theoretical foundation for logic, and it helps focus attention on *why* logical propositions are true or false. Furthermore, natural deduction is well suited for automatic proof checking by computer, and it has (along with some closely related systems) a variety of applications in computer science.

In normal English usage, the verb *infer* means to reason about some statements in order to reach a conclusion. This is an informal process, and it can run into problems due to the ambiguities of English, but the intent is to establish that the conclusion is definitely true.

Logical inference means reasoning *formally* about a set of statements in order to decide, beyond all shadow of doubt, what is true. In order to make the reasoning absolutely clear cut, we need to do several things:

- The set of statements we're reasoning about must be defined. This is called the *object language*, and a typical choice would be propositional expressions.

- The methods for inferring new facts from given information must be specified precisely. These are called the *inference rules*.

- The form of argument must be defined precisely, so that if anybody claims to have an argument that proves a fact, we can determine whether it actually is a valid argument. This defines a *metalanguage* in which proofs of statements in the object language can be written. Every step in the reasoning must be justified by an inference rule.

There are several standard ways to write down formal proofs. In this book we'll use the form which is most commonly used in computer science.

Definition 4. The notation $P_1, P_2, \ldots, P_n \vdash Q$ is called a *sequent*, and it means that if all of the propositions $P_1, P_2, \ldots, P_n$ are known, then the proposition Q can be inferred formally using the inference rules of natural deduction.

We have seen two similar notations: with truth tables, we had metalogical statements with the $\models$ operator, like $P \models Q \rightarrow P$. This statement means that if P is True, then the proposition $Q \rightarrow P$ is also True. It says nothing about how we know that to be the case. In contrast, the notation $P \vdash Q \rightarrow P$, which will be used in this section, means *there is a proof* of $Q \rightarrow P$, and the proof assumes P. Both $\models$ and $\vdash$ are used to state theorems; the distinction is that $\models$ is concerned with the ultimate truth of the propositions, while $\vdash$ is concerned with whether we have a proof. We will return to the relationship between $\models$ and $\vdash$ in Section 2.9.

In formal logic, you can't just make intuitive arguments and hope they are convincing. Every step of reasoning you make has to be backed up by an inference rule. Intuitively, an inference rule says, '*If you know that Statement1 and Statement2 are established (either assumed or proven), then you may infer Statement3,*' and furthermore, the inference constitutes a proof. *Statement1* and *Statement2* are called the *assumptions* of the inference rule, and *Statement3* is called the *conclusion*. (There may be any number of assumptions—there don't have to be just two of them.) Here's an example of an inference rule about the $\wedge$ operator, written informally in English:

> If you know some proposition a, and also the proposition b, then you are allowed to infer the proposition $a \wedge b$.

In this example, there are two assumptions—a and b—and the conclusion is $a \wedge b$. Note that we have expressed this inference using *metavariables* a and b. These metavariables could stand for any WFF; thus a might be P, $Q \rightarrow R$, etc. The variables P, Q etc, are propositional variables, belonging to the object language, and their values are True or False. We will adopt the following convention:

- Metavariables belong to the metalanguage, and are written as lower case letters $a, b, c, \ldots$. The value of a metavariable is a WFF. For example a might have the value $P \wedge Q$.

- Propositional variables belong to the object language, and are written as upper case letters $A, B, \ldots, P, Q, R, \ldots$. The value of a propositional variable is either True or False.

Formally, an inference rule is expressed by writing down the assumptions above a horizontal line, and writing the conclusion below the line:

$$\frac{\text{Statement1} \qquad \text{Statement2}}{\text{Statement3}}$$

This says that if you can somehow establish the truth of Statement1 and Statement2, then Statement3 is guaranteed (by the inference rule) to be true. The inference about $\wedge$ would be written formally as

$$\frac{a \qquad b}{a \wedge b} \ .$$

An inference rule works in only one direction—for example, if you know that P is true, you *cannot* use the rule above to infer $P \wedge Q$. After all, you don't know the value of Q, which might be False.

Figure 2.1 summarises all the inference rules of propositional logic. In the next several sections we will work systematically through them all. It would be a good idea to refer back frequently to Figure 2.1.

Many of the rules fall into two categories. The *introduction rules* have a conclusion into which a new logical operator has been introduced; they serve to build up more complex expressions from simpler ones. In contrast, the *elimination rules* require an assumption that contains a logical operator which is eliminated from the conclusion; these are used to break down complex expressions into simpler ones. Introduction rules tell you what you need to know in order to introduce a logical operator; elimination rules tell you what you can infer from a proposition containing a particular operator.

2.5.1 Definitions of True, $\neg$ and $\leftrightarrow$

Natural deduction works with a very minimal set of basic operators. In fact, the only primitive built-in objects are the constant False, and the three operators

$$\frac{a \quad b}{a \wedge b}\{\wedge I\} \qquad \frac{a \wedge b}{a}\{\wedge E_L\} \qquad \frac{a \wedge b}{b}\{\wedge E_R\}$$

$$\frac{a}{a \vee b}\{\vee I_L\} \qquad \frac{b}{a \vee b}\{\vee I_R\} \qquad \frac{a \vee b \quad a \vdash c \quad b \vdash c}{c}\{\vee E\}$$

$$\frac{a \vdash b}{a \rightarrow b}\{\rightarrow I\} \qquad \frac{a \quad a \rightarrow b}{b}\{\rightarrow E\}$$

$$\frac{a}{a}\{ID\} \qquad \frac{\text{False}}{a}\{CTR\} \qquad \frac{\neg a \vdash \text{False}}{a}\{RAA\}$$

Figure 2.1: Inference Rules of Propositional Logic.

$\wedge$, $\vee$ and $\rightarrow$. Everything else is an abbreviation! It's particularly intriguing that False is the only primitive logic value in the natural deduction system.

Definition 5. The constant True and the operators $\neg$ and $\leftrightarrow$ are abbreviations defined as follows:

$$\begin{aligned} \text{True} &= \text{False} \rightarrow \text{False} \\ \neg a &= a \rightarrow \text{False} \\ a \leftrightarrow b &= (a \rightarrow b) \wedge (b \rightarrow a) \end{aligned}$$

You can check all of these definitions semantically, using truth tables. In natural deduction, however, we will manipulate them only using the inference rules, and it will gradually become clear that the abbreviations work perfectly with the inference rules. The definition of $\leftrightarrow$ should be clear, but let's look at $\neg$ and True more closely.

Notice that $\neg$False is defined to be False $\rightarrow$ False, which happens to be the definition of True. In other words, $\neg$False $=$ True. Going the other direction, $\neg$True becomes True $\rightarrow$ False when the $\neg$ abbreviation is expanded out, and that becomes (False $\rightarrow$ False) $\rightarrow$ False when the True is expanded out. Later we will see an inference rule (called *Reductio ad Absurdum*) which will allow us to infer False from (False $\rightarrow$ False) $\rightarrow$ False. The result is that $\neg$True $=$ False and $\neg$False $=$ True.

We will write propositions with True, $\neg$ and $\leftrightarrow$ just as usual, but sometimes to make a proof go through it will be necessary to expand out the abbreviations and work just with the primitives.

2.5.2   And Introduction $\{\wedge I\}$

The *And Introduction* inference rule says that if you know that some proposition a is true, and you also know that b is true, then you may infer that the proposition $a \wedge b$ is true. As the name 'And Introduction' suggests, the rule specifies what you have to know in order to infer a proposition with a new occurrence of $\wedge$.

When we write an inference, the horizontal line will be annotated with the name of the inference rule that was used. The abbreviation $\{\wedge I\}$ stands for 'And Introduction', so the rule is written as follows:

$$\frac{a \quad b}{a \wedge b}\{\wedge I\}$$

We will now work through several examples, starting with a theorem saying that the conclusion $P \wedge Q$ can be inferred from the assumptions P and Q.

Theorem 1. $P, Q \vdash P \wedge Q$

Proof. The theorem is proved by just one application of the $\{\wedge I\}$ inference; it's the simplest possible example of how to use the $\{\wedge I\}$ inference rule.

$$\frac{P \quad Q}{P \wedge Q}\{\wedge I\}$$

$\square$

Notice that the theorem above involves two specific propositional variables, P and Q. The $\{\wedge I\}$ rule does not require any particular propositions. It uses the meta variables a and b, which can stand for any well-formed formula. For example, the following theorem has the same structure as the previous one, and is also proved with just one application of the $\{\wedge I\}$ rule, but this time the meta variable a stands for $R \to S$ and b stands for $\neg P$.

Theorem 2. $(R \to S), \neg P \vdash (R \to S) \wedge \neg P$

Proof.

$$\frac{(R \to S) \quad \neg P}{(R \to S) \wedge \neg P}\{\wedge I\}$$

$\square$

Usually you need several steps to prove a theorem, and each step requires explicit use of an inference rule. You can build up a proof in one diagram. When a subproof results in a conclusion that serves as an assumption for a larger proof, the entire subproof diagram appears above the line for the main inference in the large proof. Here is an example:

Theorem 3. $P, Q, R \vdash (P \wedge Q) \wedge R$

Proof. The main inference here has two assumptions: $P \wedge Q$ and R. However, the first assumption $P \wedge Q$ is not one of the assumptions of the entire theorem; it is the conclusion of another $\{\wedge I\}$ inference with assumptions P and Q. The entire proof is written in a single diagram. Notice how the conclusion $P \wedge Q$ of the first inference is sitting in exactly the right place above the longer line (the line belonging to the main inference).

$$\cfrac{\cfrac{P \quad Q}{P \wedge Q}{\scriptstyle\{\wedge I\}} \qquad R}{(P \wedge Q) \wedge R}{\scriptstyle\{\wedge I\}}$$

□

Inference proofs have a natural tree structure: assumptions of the theorem are like the leaves of a tree, and subproofs are like the nodes (forks) of a tree. The method we are using here for writing proofs makes this tree structure explicit.

There is an alternative format for writing logical inference proofs, where each inference is written on a numbered line. When the conclusion of one inference is required as an assumption to another one, this is indicated by explicit references to the line numbers. This format looks like a flat list of statements (much like an assembly language program), while the tree format we are using has a nested structure (much like a program in a block structured language).

Logical inference has many important applications in computer science. The tree format is normally used for computing applications, so we will use that notation in this book in order to help prepare you for more advanced studies. A clear exposition of the line-number format, which won't appear in this book, appears in Lemmon's book *Beginning Logic* [20].

Exercise 3. Prove $P, Q, R \vdash P \wedge (Q \wedge R)$.

Exercise 4. Consider the following two propositions:

$$x = A \wedge (B \wedge (C \wedge D))$$
$$y = (A \wedge B) \wedge (C \wedge D)$$

Describe the shapes of the proofs for x and y. Suppose each proposition has 2^n propositional variables. What then would be the heights of the proof trees?

2.5.3   And Elimination $\{\wedge E_L\}$, $\{\wedge E_R\}$

There are two inference rules that allow the elimination of an And operation. These rules say that if $a \wedge b$ is known to be true, then a must be true, and also b must be true. The 'And Elimination Left' $\{\wedge E_L\}$ rule retains the left argument of $a \wedge b$ in the result, while the 'And Elimination Right' $\{\wedge E_R\}$ rule retains the right argument.

$$\frac{a \wedge b}{a}\{\wedge E_L\} \qquad \frac{a \wedge b}{b}\{\wedge E_R\}$$

Here is a simple example using And Elimination:

Theorem 4. $P, Q \wedge R \vdash P \wedge Q$

Proof. Two inferences are required. First the 'Left' form of the And Elimination rule $\{\wedge E_L\}$ is used to obtain the intermediate result Q, and the And Introduction rule is then used to combine this with P to obtain the result $P \wedge Q$.

$$\frac{P \qquad \dfrac{Q \wedge R}{Q}\{\wedge E_L\}}{P \wedge Q}\{\wedge I\}$$

□

The next example contains several applications of And Elimination, including both the Left and Right forms of the rule. As these two examples show, the three rules $\{\wedge I\}$, $\{\wedge E_L\}$ and $\{\wedge E_R\}$ can be used systematically to deduce consequences of logical conjunctions.

Theorem 5. $(P \wedge Q) \wedge R \vdash R \wedge Q$

Proof.

$$\frac{\dfrac{(P \wedge Q) \wedge R}{R}\{\wedge E_R\} \qquad \dfrac{\dfrac{(P \wedge Q) \wedge R}{P \wedge Q}\{\wedge E_L\}}{Q}\{\wedge E_R\}}{R \wedge Q}\{\wedge I\}$$

□

One of the most fundamental properties of the $\wedge$ operator is that it is *commutative*: the proposition $P \wedge Q$ is equivalent to $Q \wedge P$. Later we will prove that, but for now we just consider the following simpler theorem.

Theorem 6. $P \wedge Q \vdash Q \wedge P$

Proof. The idea behind this proof is to infer Q and P—in that order, with Q to the left of P—above the main line, so $\{\wedge I\}$ can be used to infer $Q \wedge P$. The intermediate results Q and P are obtained by And Elimination.

$$\frac{\dfrac{P \wedge Q}{Q}\{\wedge E_R\} \qquad \dfrac{P \wedge Q}{P}\{\wedge E_L\}}{Q \wedge P}\{\wedge I\}$$

□

Inference proof trees can become quite large, but you can always work through them systematically, one inference at a time, until you see how all the parts of the proof fit together. You can check the individual inferences in any order you like, either top-down or bottom-up or any other order you like. We will give one more example of a theorem proved with the And Introduction and Elimination rules, with a larger proof.

Theorem 7. For any well formed propositions a, b and c,

$$a \wedge (b \wedge c) \vdash (a \wedge b) \wedge c$$

Proof.

$$\frac{\dfrac{\dfrac{a \wedge (b \wedge c)}{a}\{\wedge E_L\} \quad \dfrac{\dfrac{a \wedge (b \wedge c)}{b \wedge c}\{\wedge E_R\}}{b}\{\wedge E_L\}}{a \wedge b}\{\wedge I\} \quad \dfrac{\dfrac{a \wedge (b \wedge c)}{b \wedge c}\{\wedge E_R\}}{c}\{\wedge E_R\}}{(a \wedge b) \wedge c}\{\wedge I\}$$

□

Exercise 5. Prove $(P \wedge Q) \wedge R \vdash P \wedge (Q \wedge R)$.

2.5.4 Imply Elimination $\{\to E\}$

As we work through the inference rules, you should refer back frequently to Figure 2.1. This will help build up a coherent picture of the complete system of natural deduction. Although the rules relating to $\vee$ come next in the figure, we will first study the rules for $\to$, which are slightly simpler.

The Imply Elimination rule $\{\to E\}$ says that if you know a is true, and also that it implies b, then you can infer b. The traditional Latin name for the rule is *Modus Ponens*.

$$\boxed{\dfrac{a \qquad a \to b}{b}\{\to E\}}$$

The following theorem provides a simple example of the application of $\{\to E\}$.

Theorem 8. $Q \wedge P,\ P \wedge Q \to R \vdash R.$

Proof.

$$\dfrac{\dfrac{\dfrac{Q \wedge P}{P}\{\wedge E_R\} \quad \dfrac{Q \wedge P}{Q}\{\wedge E_L\}}{P \wedge Q}\{\wedge I\} \quad P \wedge Q \to R}{R}\{\to E\}$$

□

Often we have chains of implication of the form $a \to b,\ b \to c$ and so on. The following theorem says that given a and these linked implications, you can infer c.

Theorem 9. For all propositions a, b and c, $a, a \to b, b \to c \vdash c.$

Proof.

$$\dfrac{\dfrac{a \quad a \to b}{b}\{\to E\} \quad b \to c}{c}\{\to E\}$$

□

Exercise 6. Prove $P,\ P \to Q,\ (P \wedge Q) \to (R \wedge S) \vdash S.$

Exercise 7. Prove $P \to Q,\ R \to S,\ P \wedge R \vdash S \wedge R.$

2.5.5 Imply Introduction $\{\to I\}$

The Imply Introduction rule $\{\to I\}$ says that, in order to infer the logical implication $a \to b$, you must have a proof of b using a as an assumption.

$$\boxed{\dfrac{a \vdash b}{a \to b}\{\to I\}}$$

We will first give a simple example of Imply Introduction, and then discuss the important issue of keeping track of the assumptions that have been made.

Theorem 10. $\vdash (P \wedge Q) \to P.$

Proof. First consider the sequent $P \wedge Q \vdash Q$, which is proved by the following And Elimination inference:

$$\frac{P \wedge Q}{Q}\{\wedge E_R\}$$

Now we can use the sequent (that is, the theorem established by the inference) and the $\{\rightarrow I\}$ rule:

$$\frac{P \wedge Q \vdash Q}{P \wedge Q \rightarrow Q}\{\rightarrow I\}$$

$\square$

It is crucially important to be careful about what we are assuming. In fact, the reason for having the sequent notation with the $\vdash$ operator is to write down the assumptions of a theorem just as precisely as the conclusion.

In the $\{\wedge E_R\}$ inference we assumed $P \wedge Q$; without that assumption we could not have inferred Q. Therefore this assumption appears in the sequent to the left of the $\vdash$. However, the entire sequent $P \wedge Q \vdash Q$ does not rely on any further assumptions; it is independently true. Therefore we can put it above the line of the $\{\rightarrow I\}$ rule 'for free,' without actually making any assumptions. Since nothing at all needed to be assumed to support the application of the $\{\rightarrow I\}$ rule, we end up with a theorem that doesn't require any assumptions. The sequent that expresses the theorem, $\vdash P \wedge Q \rightarrow P$, has nothing to the left of the $\vdash$ operator.

It is customary to write the entire proof as a single tree, where a complete proof tree appears above the line of the $\{\rightarrow I\}$ rule. That allows us to prove $\vdash P \wedge Q \rightarrow P$ with just one diagram:

$$\frac{\dfrac{P \wedge Q}{P}\{\wedge E_R\}}{P \wedge Q \rightarrow P}\{\rightarrow I\}$$

From this diagram, it looks like there is an assumption $P \wedge Q$. However, that was a temporary, local assumption whose only purpose was to establish $P \wedge Q \vdash P$. Once that result is obtained the assumption $P \wedge Q$ can be thrown away. An assumption that is made temporarily only in order to establish a sequent, and which is then thrown away, is said to be *discharged*. A discharged assumption does *not* need to appear to the left of the $\vdash$ of the main theorem. In our example, the proposition $P \wedge Q \rightarrow P$ is *always* true, and it doesn't matter whether $P \wedge Q$ is true or false.

In big proof trees it may be tricky to keep track of which assumptions have been discharged and which have not. We will indicate the discharged assumptions by putting a box around them. (A more common notation is to draw a line through the discharged assumption, but for certain propositions that leads to a highly unreadable result.) Following this convention, the proof becomes:

$$\frac{\dfrac{\boxed{P \wedge Q}}{P}\,{\scriptstyle\{\wedge E_R\}}}{P \wedge Q \to P}\,{\scriptstyle\{\to I\}}$$

Recall the proof of Theorem 9, which used the chain of implications $a \to b$ and $b \to c$ to infer c, given also that a is true. A more satisfactory theorem would just focus on the chain of implications, without relying on a actually being true. The following theorem gives this purified chain property: it says that if $a \to b$ and $b \to c$, then $a \to c$.

Theorem 11 (Implication chain rule). For all propositions a, b and c,

$$a \to b,\ b \to c \vdash a \to c$$

Proof. Since we are proving an implication $a \to c$, we need to use the $\{\to I\}$ rule—there is no other way to introduce the $\to$ operator! That rule requires a proof of c given the assumption a, which is essentially the same proof tree used before in Theorem 9. The important point, however, is that the assumption a is discharged when we apply $\{\to I\}$. The other two assumptions, ($a \to b$ and $b \to c$), are not discharged. Consequently a doesn't appear to the left of the $\vdash$ in the theorem; instead it appears to the left of the $\to$ in the *conclusion* of the theorem.

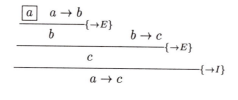

$\square$

Sometimes in large proofs it can be confusing to see just where an assumption has been discharged. You may have an assumption P with a box around it, but there could be all sorts of implications $P \to \cdots$ which came from somewhere else. In such cases it's probably clearer to build up the proof in stages, with several separate proof tree diagrams.

A corollary of this implication chain rule is the following theorem, which says that if $a \to b$ but you know that b is false, than a must also be false. This is an important theorem which is widely used to prove other theorems, and its traditional Latin name is *Modus Tollens*.

Theorem 12 (*Modus Tollens*). For all propositions a and b,

$$a \to b,\ \neg b \vdash \neg a$$

Proof. First we need to expand out the abbreviations, using the definition that $\neg a$ means $a \to$ False. This results in the following sequent to be proved: $a \to b,\ b \to$ False $\vdash a \to$ False. This is an instance of Theorem 11. □

Exercise 8. Prove $P \vdash Q \to P \wedge Q$.

Exercise 9. Prove $\vdash P \wedge Q \to Q \wedge P$.

2.5.6 Or Introduction $\{\vee I_L\},\ \{\vee I_R\}$

Like all the introduction rules, Or Introduction specifies what you have to establish in order to infer a proposition containing a new $\vee$ operator. If the proposition a is true, then both $a \vee b$ and $b \vee a$ must also be true (you can see this by checking the truth table definition of $\vee$). Or Introduction comes in two forms, Left and Right.

$$\frac{a}{a \vee b}\{\vee I_L\} \qquad \frac{b}{a \vee b}\{\vee I_R\}$$

Theorem 13. $P \wedge Q \vdash P \vee Q$

Proof. The proof requires the use of Or Introduction. There are two equally valid ways to organise the proof. One method begins by establishing P, and then uses $\{\vee I_L\}$ to put the P to the left of $\vee$:

$$\frac{\dfrac{P \wedge Q}{P}\{\wedge E_L\}}{P \vee Q}\{\vee I_L\}$$

An alternative proof first establishes Q and then uses $\{\vee I_R\}$ to put the Q to the right of $\vee$:

$$\frac{\dfrac{P \wedge Q}{Q}\{\wedge E_R\}}{P \vee Q}\{\vee I_R\}$$

Normally, of course, you would choose one of these proofs randomly; there is no reason to give both. □

Exercise 10. Prove $\vdash P \to$ False $\vee P$.

Exercise 11. Prove $P,\ Q \vdash (P \wedge Q) \vee (Q \vee R)$.

2.5.7   Or Elimination $\{\vee E\}$

The Or Elimination rule specifies what you can conclude if you know that a proposition of the form $a \vee b$ is true. We can't conclude anything about either a or b, since either of those might be false even if $a \vee b$ is true. However, suppose we know $a \vee b$ is true, and also suppose there is some conclusion c that can be inferred from a and can also be inferred from b. Then c must also be true.

Or Elimination is a formal version of *proof by case analysis*. It amounts to the following argument: 'There are two cases: (1) if a is true, then c holds; (2) if b is true then c holds. Therefore c is true'. Here is an example:

Theorem 14. $(P \wedge Q) \vee (P \wedge R) \vdash P$

Proof. There are two proofs above the line. The first proof assumes $P \wedge Q$ in order to infer P. However, that inference is all that we need; $P \wedge Q$ is discharged and is not an assumption of the main theorem. For the same reason, $P \wedge R$ is also discharged. The only undischarged assumption is $(P \wedge Q) \vee (P \vee R)$, which therefore must appear to the left of the $\vdash$ in the main theorem.

$$
\cfrac{(P \wedge Q) \vee (P \wedge R) \qquad \cfrac{\boxed{P \wedge Q}}{P}{\{\wedge E_L\}} \qquad \cfrac{\boxed{P \wedge R}}{P}{\{\wedge E_L\}}}{P}\{\vee E\}
$$

$\square$

Finally, here is a slightly more complex example:

Theorem 15. $(a \wedge b) \vee (a \wedge c) \vdash b \vee c$

Proof.

$$
\cfrac{(a \wedge b) \vee (a \wedge c) \qquad \cfrac{\cfrac{\boxed{a \wedge b}}{b}\{\wedge E_R\}}{b \vee c}\{\vee I_L\} \qquad \cfrac{\cfrac{\boxed{a \wedge c}}{c}\{\wedge E_R\}}{b \vee c}\{\vee I_R\}}{b \vee c}\{\vee E\}
$$

$\square$

2.5.8 Identity {*ID*}

The Identity rule {*ID*} says, rather obviously, that if you know a is true, then you know a is true.

$$\boxed{\dfrac{a}{a}\{ID\}}$$

Although the Identity rule seems trivial, it is also necessary. Without it, for example, there would be no way to prove the following theorem, which we would certainly want to be true.

Theorem 16. $P \vdash P$

Proof.

$$\frac{P}{P}\{ID\}$$

$\square$

An interesting consequence of the Identity rule is that True is true, as the following theorem states.

Theorem 17. $\vdash$ True

Proof. Recall that True is an abbreviation for False $\rightarrow$ False. We need to infer this implication using the {$\rightarrow I$} rule, and that in turn requires a proof of False given the assumption False. This can be done with the {*ID*} rule; it could also be done with the Contradiction rule, which we'll study shortly.

$$\frac{\dfrac{\boxed{\text{False}}}{\text{False}}\{ID\}}{\text{False} \rightarrow \text{False}}\{\rightarrow I\}$$

Notice that we had to assume False in order to prove this theorem, but fortunately that assumption was discharged when we inferred False $\rightarrow$ False. The assumption of False was just temporary, and it doesn't appear as an assumption of the theorem. That's a relief; it would never do if we had to assume that False is true in order to prove that True is true! $\square$

2.5.9 Contradiction {*CTR*}

The Contradiction rule says that you can infer *anything at all* given the assumption that False is true.

$$\frac{\mathsf{False}}{a}\{CTR\}$$

In effect, this rule says that False is untrue, and it expresses that fact purely through the mechanism of logical inference. It would be disappointing if we had to describe the fundamental falseness of False by making a meaningless statement outside the system, such as 'False is wrong.' After all, the whole point of natural deduction is to describe the process of logical reasoning formally, using a small set of clearly specified inference rules. It would also be a bad idea to try to define False as equal to ¬True. Since True is already defined to be ¬False, that would be a meaningless and useless circular definition.

The Contradiction rule describes the untruthfulness of False indirectly, by saying that everything would become provable if False is ever assumed or inferred.

Theorem 18. $P, \neg P \vdash Q$

Proof. Recall that $\neg P$ is just an abbreviation for $P \rightarrow$ False. That means we can use the $\{\rightarrow E\}$ rule to infer False, and once that happens we can use $\{CTR\}$ to support any conclusion we feel like—even Q, which isn't even mentioned in the theorem's assumptions!

$$\frac{\dfrac{P \qquad P \rightarrow \mathsf{False}}{\mathsf{False}}\{\rightarrow E\}}{Q}\{CTR\}$$

$\square$

The Identity and Contradiction rules often turn up in larger proofs. A typical example occurs in the following theorem, which states an important property of the logical Or operator $\vee$. This theorem says that if $a \vee b$ is true, but a is false, then b has to be true. It should be intuitively obvious, but the proof is subtle and merits careful study.

Theorem 19. For all propositions a and b,

$$a \vee b, \neg a \vdash b$$

Proof. As usual, the $\neg a$ abbreviation should be expanded to $a \rightarrow$ False. Since we're given $a \vee b$, the basic structure will be an Or Elimination, and there will be two smaller proof trees corresponding to the two cases for $a \vee b$. In the first case, when a is assumed temporarily, we obtain a contradiction with the theorem's assumption of $\neg a$: False is inferred, from which we can infer anything else (and here we want b). In the second case, when b is assumed temporarily, the desired result of b is an immediate consequence of the $\{ID\}$ rule.

Note that there are two undischarged assumptions, $a \vee b$ and $a \rightarrow$ False, and there are two temporary assumptions that are discharged by the $\{\vee E\}$ rule. □

2.5.10 *Reductio ad Absurdum* $\{RAA\}$

The *Reductio ad Absurdum* (reduce to absurdity) rule says that if you can infer False from an assumption $\neg a$, then a must be true. This rule underpins the *proof by contradiction* strategy: if you want to prove a, first assume the contradiction $\neg a$ and infer False; the $\{RAA\}$ rule then allows you to infer a.

$$
\frac{\neg a \vdash \text{False}}{a} \; \{RAA\}
$$

Theorem 20 (Double negation). $\neg\neg a \vdash a$

Proof. Our strategy is to use a proof by contradiction. That is, if we can assume $\neg a$ and infer False, the $\{RAA\}$ rule would then yield the inference a. Since we are given $\neg\neg a$, the contradiction will have the following general form:

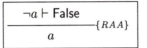

To make this go through, we need to replace the abbreviations by their full defined values. Recall that $\neg a$ is an abbreviation for $a \rightarrow$ False, so $\neg\neg a$ actually means $(a \rightarrow \text{False}) \rightarrow \text{False}$. Once we expand out these definitions, the inference becomes much clearer: it is just a simple application of $\{\rightarrow E\}$:

$$
\frac{\dfrac{\boxed{a \rightarrow \text{False}} \quad (a \rightarrow \text{False}) \rightarrow \text{False}}{\text{False}} \; \{\rightarrow E\}}{a} \; \{RAA\}
$$

Both requirements for the $\{RAA\}$ rule have now been provided, and the $\{RAA\}$ gives the final result a. □

2.5.11   Inferring the Operator Truth Tables

The inference rules of natural deduction are intended to serve as a formal foundation for logical reasoning. This raises two fundamental questions: *Are the inference rules actually powerful enough? Will they ever allow us to make mistakes?* These are profound and difficult questions. Modern research in mathematical logic addresses such questions, and has led to some astonishing results. In this section we start with an easy version of the first question: *Are the inference rules powerful enough to calculate the truth tables of the logical operators?*

Answering this question will provide some good practice in using the inference rules. A deeper point is that it's philosophically satisfying to know that the inference rules provide a complete foundation for propositional logic. Nothing else is required: the truth tables given earlier in the chapter are the most intuitive place to start learning logic, and we treated them like definitions at the beginning of the chapter, but it isn't necessary to accept them as the fundamental definitions of the operators. If we take instead the inference rules as the foundation of mathematical logic, then truth tables are no longer definitions; they merely summarise a lot of calculations using the rules.

To illustrate the idea, let's use the inference rules to calculate the value of True $\wedge$ False. To answer this fully, we need to prove that True $\wedge$ False is logically equivalent to False, and also that it is *not* logically equivalent to True. Recall that False is a primitive constant, but True is defined as False $\rightarrow$ False. First, here is a proof of the sequent $\vdash$ True $\wedge$ False $\rightarrow$ False:

$$\frac{\dfrac{(\text{False} \rightarrow \text{False}) \wedge \text{False}}{\text{False}}\{\wedge E_R\}}{((\text{False} \rightarrow \text{False}) \wedge \text{False}) \rightarrow \text{False}}\{\rightarrow I\}$$

Next, we prove the sequent $\vdash$ False $\rightarrow$ ((False $\rightarrow$ False) $\wedge$ False):

$$\frac{\dfrac{\text{False}}{(\text{False} \rightarrow \text{False}) \wedge \text{False}}\{CTR\}}{\text{False} \rightarrow ((\text{False} \rightarrow \text{False}) \wedge \text{False})}\{\rightarrow I\}$$

Putting these results together, and reintroducing the abbreviations, we get

$$\text{True} \wedge \text{False} \leftrightarrow \text{False}$$

We have thereby calculated one of the lines of the truth table definition of $\wedge$. The complete truth tables for all the logical operators can be inferred using similar calculations.

Exercise 12. Use the inference rules to calculate the value of True $\wedge$ True.

Exercise 13. Use the inference rules to calculate the value of True $\vee$ False.

Exercise 14. Notice that in the proof of ⊢ True∧False → False we used {∧E_R} to obtain False from (False → False) ∧ False, and everything worked fine. However, we could have used {∧E_L} instead to infer False → False, which is True. What would happen if that choice is made? Would it result in calculating the wrong value of True ∧ False? Is it possible to show that True ∧ False is *not* logically equivalent to True?

2.6 Proof Checking by Computer

One of the great benefits of formal logic is the possibility of using computer software to check proofs automatically. Informal arguments in English are full of imprecision and ambiguity, so there is no hope of writing a computer program to determine whether one is valid. People don't always agree as to whether an argument holds water! Formal proofs, however, are totally precise and unambiguous. Formal arguments may become long and involved, but computers are good at long and involved calculations.

Formal proofs are intended to provide the utmost confidence that a theorem is correct. The language of propositional logic, along with the natural deduction inference system, are solid foundations for accurate reasoning. However, everyone makes mistakes, and even a tiny error would make a large proof completely untrustworthy. To get the full benefit of formal logic, we need computers to help with the proofs.

A *proof checker* is a computer program that reads in a theorem and a proof, and determines whether the proof is valid and actually establishes the theorem. A *theorem prover* is a computer program that reads in a theorem and attempts to generate a proof from scratch. The generic term *proof tools* refers to any computer software that helps with formal proofs, including proof checkers and theorem provers.

The advantage of a theorem prover is that it can sometimes save the user a lot of work. However, theorem provers don't always succeed, since there are plenty of true theorems whose proofs are too difficult for them to find. A theorem prover may stop with the message 'I cannot prove it', or it may go into an infinite loop and never stop at all. The main advantage of a proof checker is that it can *always* determine whether a purported proof is valid or invalid. Proof checkers are also suitable when you want to write a proof by hand, and you also want to be sure it's correct.

Proof tools are the subject of active current research, but they are not limited to research: they are also becoming practical for real applications. There are a number of 'industrial strength' proof tools for various logical systems (see Section 2.10 for references). Proof tools have already been applied successfully to some very difficult and important real-world problems. A recent example is the proof that the floating point hardware in the Intel Pentium Pro processor is correct [23]. An active current research topic is to find methods for making proof checkers easier to use and more helpful in real applications.

As proof checkers become more widely used, it will become increasingly important for computing professionals to have a good working understanding of logic.

We'll now look at a simple proof checking system implemented in Haskell, which is part of the software that accompanies this book. You can obtain the software and its documentation from this book's web page; see the Preface for the web address.

Studying the proof checker will introduce you to a technology that is likely to become increasingly important in the near future. You can also use the software to check your own proofs; it's nice to be sure your exercises are correct before handing them in! Furthermore, you may find it interesting to study the implementation of the checker.

Documentation of the proof checker is available on the book web page. The program is written in Haskell. Although it uses some advanced features of Haskell that aren't covered in Chapter 1, most of it isn't too difficult to understand. Haskell is well suited for writing proof checkers; indeed, most of the industrial strength proof tools are implemented in functional programming languages, since conventional languages just aren't powerful enough.

2.6.1   Example of Proof Checking

Before getting into picky details, let's start by checking a real theorem:

Theorem 21. $\vdash Q \rightarrow ((P \wedge R) \rightarrow (R \wedge Q))$

Proof. See Figure 2.3.   □

In order to process this theorem with a Haskell program, we have to represent it in a form that can be typed into a file. Figure 2.2 gives the theorem in both forms, with the mathematical notation and the proof checker's notation shown side by side. The value named `example_theorem` has type `Theorem`, and it consists of three parts: (1) a constructor `Theorem` that starts the data structure; (2) a list of assumptions (i.e. propositions to the left of the ⊢) which is [] for this example; and (3) the proposition to be proved. As you can see, the computer-readable proposition is expressed with prefix operators; the name `And` represents ∧, while the name `Imp` represents →.

Figure 2.3 shows the proof of Theorem 21. The conventional mathematical notation is given alongside the Haskell representation, so you can compare them easily. The Haskell representation of the proof is a value named `proof1`, and its type is `Proof`. Notice that the mathematical notation has a tree structure, and the proof is also represented as a tree structure in Haskell. The only difference is that the mathematical notation uses lines and positioning on the page to show the structure, while the Haskell notation uses punctuation. The computer representation uses prefix operators to indicate inferences; for example, the name `AndI` indicates the beginning of an And-Introduction $\{\wedge I\}$ inference. The proof is a value named `proof1`, and it has type `Proof`.

Theorem 21. $\vdash Q \to ((P \land R) \to (R \land Q))$

```
example_theorem :: Theorem
example_theorem =
  Theorem
    []
    (Imp Q (Imp (And P R) (And R Q)))
```

Figure 2.2: Theorem 21 and its Haskell Representation

The proof checker is a Haskell function named `check_proof` which takes two arguments: a `Theorem` and a `Proof`. The function checks to see whether the proof is valid, and also whether it serves to establish the truth of the theorem. You might expect the checker to return a result of type `Bool`, where `True` means the proof is valid and `False` means the proof is incorrect. However, it's much better for the checker to give a detailed explanation if something is wrong with the proof, and to do that it needs to perform some Input/Output. Because of this, the function returns a Haskell value of type `IO ()` instead of a simple `Bool`. You do *not* need to know how the Input/Output works. The function's type signature is:

```
check_proof :: Theorem -> Proof -> IO ()
```

The proof given in Figure 2.3 is valid: all of the inferences are sound; the conclusion of the main inference matches the result to the right of the $\vdash$ in the theorem, and there are no undischarged assumptions that fail to appear to the left of the $\vdash$ in the statement of the theorem.

To check the proof, we must first start an interactive session with Haskell and load the software tools (see the Appendix for instructions on how to do this). Here is the output produced by running the proof checker on the example:

```
> check_proof example_theorem proof1
The proof is valid
```

In order to see what happens when something goes wrong with a proof, let's introduce a small mistake into the previous example. Figure 2.4 gives the modified proof, along with its Haskell representation. The error is that the two subproofs above the line of the And Introduction step now appear in the wrong order: on the left is the assumption Q, and on the right is the proof of R. Because of this, the $\{\land I\}$ rule infers the proposition $Q \land R$, but we still have $R \land Q$ below the line. Here is the result of running the incorrect proof through the proof checker:

```
> check_proof example_theorem proof2
Invalid And-Introduction: the conclusion
```

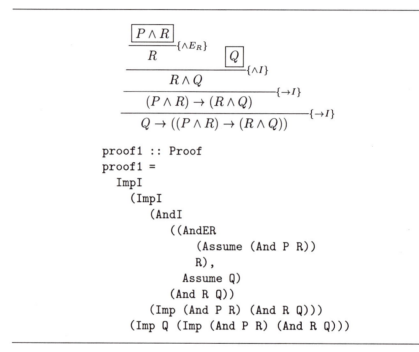

$$\cfrac{\cfrac{\boxed{P \wedge R}}{R}{\scriptstyle \{\wedge E_R\}} \quad \boxed{Q}}{\cfrac{\cfrac{R \wedge Q}{(P \wedge R) \to (R \wedge Q)}{\scriptstyle \{\to I\}}}{Q \to ((P \wedge R) \to (R \wedge Q))}{\scriptstyle \{\to I\}}}{\scriptstyle \{\wedge I\}}$$

```
proof1 :: Proof
proof1 =
  ImpI
    (ImpI
      (AndI
        ((AndER
            (Assume (And P R))
            R),
          Assume Q)
        (And R Q))
      (Imp (And P R) (And R Q)))
    (Imp Q (Imp (And P R) (And R Q)))
```

Figure 2.3: A Valid Proof of Theorem 21 and its Haskell Representation

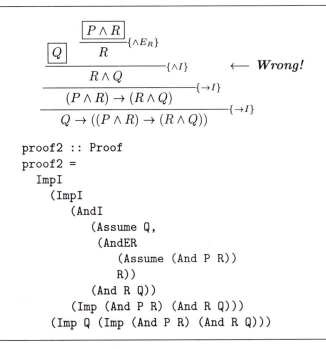

```
proof2 :: Proof
proof2 =
  ImpI
    (ImpI
      (AndI
        (Assume Q,
        (AndER
          (Assume (And P R))
          R))
        (And R Q))
      (Imp (And P R) (And R Q)))
    (Imp Q (Imp (And P R) (And R Q)))
```

Figure 2.4: An Invalid Proof of Theorem 21 and its Haskell Representation

```
    And R Q
  must be the logical And of the assumption
    Q
  with the assumption
    R
The proof is invalid
```

To use the proof checker on your own, you will need to know how to represent propositions (WFFs) and how to represent proofs. The following sections describe these issues briefly, but you should also read the online documentation on the book's web page.

Exercise 15. Suppose we simply replace $R \wedge Q$ below the $\{\wedge I\}$ line with $Q \wedge R$. This fixes the `Invalid And-Introduction` error, but it introduces another error into the proof.

(a) Edit `proof2` to reflect this change; call the result `proof3`.

(b) Decide exactly what is wrong with `proof3`.

(c) Run the proof checker on `proof3`, and see whether it reports the same error that you predicted.

2.6.2   Representation of WFFs

The well-formed formulas of propositional logic can be represented in Haskell
as an algebraic data type. This representation allows us to use the compiler
to check that a formula is indeed well-formed, and it also provides a way to
express terms that appear in logical proofs—later in this chapter we will exploit
that in order to support a program that checks logical proofs for correctness.

The Boolean constants are represented by FALSE and TRUE. Intuitively, these
correspond in meaning to the familiar Boolean constants False and True. Use
False (or True) when you want to write Boolean expressions to be evaluated;
use FALSE (or TRUE) when you want to write a WFF and reason about it. We
will come back to this subtle but important distinction at the end of the section.

The traditional names used for propositional variables are upper case letters
$P, Q, R \ldots$. We allow every upper case letter to be used as a propositional
variable, except that F and T are disallowed because someone reading the code
might wonder whether these are supposed to be variables or the constants *false*
and *true*. In addition, you can make any string into a propositional variable
by writing Pvar "name".

The logical expression $P \wedge Q$ is written And P Q. If the arguments to the And
are themselves logical expressions, they should be surrounded by parentheses,
for example: And (And P Q) R. In a similar way, $P \vee Q$ is written Or P Q,
and $P \rightarrow Q$ is written Imp P Q, and the negation $\neg P$ is written Not P. In
all cases, arguments that are not simple logical variables should be enclosed in
parentheses.

WFFs are defined directly as a Haskell data type Prop. It is common in
mathematics to define structures recursively, as WFFs were defined in the previous section. Haskell's algebraic data types allow such standard mathematical
definitions to be turned directly into a computer program. The definition is:

```
data Prop
  = FALSE
  | TRUE
  | A | B | C | D | E | G | H | I | J | K | L | M
  | N | O | P | Q | R | S | U | V | W | X | Y | Z
  | Pvar String
  | And Prop Prop
  | Or Prop Prop
  | Not Prop
  | Imp Prop Prop
  | Equ Prop Prop
  deriving (Eq,Show)
```

Exercise 16. Define each of the following well-formed formulas as a Haskell
    value of type WFF.

   (a) P

(b) $Q \lor$ False

(c) $Q \to (P \to (P \land Q))$

(d) $P \land (\neg Q)$

(e) $\neg P \to Q$

(f) $(P \land \neg Q) \lor (\neg P \land Q) \to (P \lor Q)$

Exercise 17. Translate each of the following Haskell expressions into the conventional mathematical notation.

(a) `And P Q`

(b) `Imply (Not P) (Or R S)`

(c) `Equ (Imply P Q) (Or (Not P) Q)`

2.6.3 Representing Proofs

A proof is represented by another Haskell algebraic data type. Technically, a proof is a data structure which contains a proposition along with a formal argument for the truth of the proposition. There is a separate constructor for every kind of proof:

- `Assume Prop`. The simplest way to establish that a proposition is true is to assume it. No further justification is necessary, and any proposition at all may be assumed. See `proof1` above for an example. Note that assumptions may be discharged by the Imply Introduction rule $\{\to I\}$, so there is a limited and well-defined scope for all assumptions. Unfortunately, however, it isn't easy to see what the scope of an assumption is just by looking at it; you need to study what it means to discharge.

- `AndI (Proof,Proof) Prop`. The And Introduction inference rule $\{\land I\}$ requires two separate proofs above the line, as well as the claimed conclusion (which is a proposition).

- `AndEL Proof Prop`. The And Elimination Left inference rule $\{\land E_L\}$ has just one proof above the line; like all the inference rules it has exactly one conclusion, which is a proposition.

- `AndER Proof Prop`. The And Elimination Right rule $\{\land E_R\}$. This is the Right version of And Elimination $\{\land E_R\}$

- `OrIL Proof Prop`. The Or Introduction Left rule $\{\lor I_L\}$;

- `OrIR Proof Prop`. The Or Introduction Right rule $\{\lor I_R\}$;

- OrE (Proof,Proof,Proof) Prop. The Or Elimination rule $\{\vee E\}$ requires three proofs above the line. The first one is an application of the form $a \vee b$, the second is a proof of some conclusion c given a, and the third is a proof of the same conclusion c given b. The conclusion of the rule must be the proposition c.

- ImpI Proof Prop. The Imply Introduction rule $\{\to I\}$.

- ImpE (Proof,Proof) Prop. The Imply Elimination rule $\{\to E\}$.

- ID Proof Prop. The Identity rule $\{ID\}$.

- CTR Proof Prop. The Contradiction rule $\{CTR\}$.

- RAA Proof Prop. The *Reductio ad Absurdum* rule $\{RAA\}$.

The best way to understand how to represent proofs is to look at some examples, and then try your own. Start by studying **proof1** and **proof2** above.

The representation of a theorem is very simple: it just contains a list of assumptions of type [Prop] and a single conclusion of type Prop.

```
data Theorem = Theorem [Prop] Prop
```

To represent a theorem of the form

$$a_1, a_2, a_3 \vdash c,$$

you would write

$$\text{Theorem } [a_1, \ a_2, \ a_3] \ c.$$

The proof checker function takes two arguments: a theorem of type Theorem and a proof of type Proof. The easiest way to use it is to give a name to the theorem and the proof, write a defining equation for each, and save the definitions in a file. Don't try to type in proofs interactively! Follow the defining equations above for example_theorem and proof1 as a model.

There are many things that can go wrong with a proof. The proof checker tries to give the best messages possible, but sometimes it is hard to debug a proof. Actually, proof debugging has much in common with program debugging. See the book's web page for hints on debugging proofs.

2.7   Boolean Algebra: Equational Reasoning

We have already looked at two major approaches to propositional logic: the *semantic* approach with truth tables, and the *syntactic* approach with the inference rules of natural deduction. We now look at the third major system, Boolean algebra, which is an *axiomatic* approach to logic.

The earliest attempts to develop a formal logic system were based on inference. The most famous of these was Aristotle's Theory of Syllogisms, which profoundly affected the entire development of logic and philosophy for more than two thousand years.

During the last several centuries, however, a completely different style of mathematical reasoning appeared: algebra. Algebraic techniques were enormously successful for reasoning about numbers, polynomials and functions. One of the most appealing benefits of algebra is that many problems can be expressed as an equation involving an unknown quantity x that you would like to determine; you can then use algebraic laws to manipulate the equation systematically in order to solve for x.

A natural question is: can the power of algebraic techniques be applied to other areas of mathematics? Perhaps the most famous such application is Descartes' analytical geometry. George Boole, a nineteenth century British mathematician, saw that algebraic methods might also be applied to make formal logical reasoning easier, and he attempted to develop such a system. Boole's approach had some technical flaws, and is no longer in use. However, a successful modern algebraic approach to logic has been developed and is named in his honour.

Boolean algebra is a form of *equational reasoning*. There are two crucial ideas: (1) you show that two values are the same by building up *chains of equalities*, and (2) you can *substitute equals for equals* in order to add a new link to the chain.

A *chain of equalities* relies on the fact that if you know $a = b$ and also $b = c$, then you can deduce formally that $a = c$. For example, the following chain of equations allows us to conclude that $a = e$:

$$
\begin{aligned}
a &= b \\
  &= c \\
  &= d \\
  &= e
\end{aligned}
$$

Substituting equals for equals means that if you know $x = y$, and if you have a big expression which contains x, then you can replace x by y without changing the value of the big expression. For example, suppose you're given that $x = 2 + p$ and $y = 5 \times x + 3$. Then you replace x by $2 + p$, resulting in $y = 5 \times (2 + p) + 3$.

There is a minor but important point to observe in the example above: you have to use parentheses properly to ensure that the value you substitute into the expression sticks together as one value. In this example, we had to put parentheses around the value $2 + p$, because otherwise we would have written the incorrect equation $y = 5 \times 2 + p + 3$, which has a quite different meaning.

When we build a chain of equations using Boolean algebra, it's good practice to give a justification for each step in the chain. The justifications help a reader to understand the proof, and they also make it easier to check that the proof is actually valid. A standard way to write chains of equations is to start each line (except the first) with an $=$ sign, followed by the next expression in our chain, followed by the justification which explains how we obtained it from the previous expression. You'll see plenty of examples as we work through the laws of Boolean algebra.

2.7.1   The Laws of Boolean Algebra

Modern Boolean algebra is based on a set of equations that describe the basic algebraic properties of propositions. These equations are called *laws*; a law is a proposition that is always true, for every possible assignment of truth values to the logical variables.

The laws of Boolean Algebra are analogous to ordinary algebraic laws. For example, elementary algebra has commutative equations for addition and multiplication, $x + y = y + x$ and $x \times y = y \times x$. There are analogous commutative laws for $\vee$ and $\wedge$, saying that $x \vee y = y \vee x$ and $x \wedge y = y \wedge x$. It can be enlightening to compare the laws of Boolean algebra with those of elementary algebra, but don't get carried away: there are differences as well as similarities.

There is one particularly dangerous trap. In many ways, logical And ($\wedge$) behaves like multiplication ($\times$) while logical Or ($\vee$) behaves like addition ($+$). In fact, these similarities have tempted many people to use the $+$ symbol for Or and the $\times$ symbol (or $\cdot$) for And. George Boole carried similar analogies very far—much too far—in his original work.

However, $\wedge$ does not behave like $\times$ in all respects, and $\vee$ does not behave like $+$ in all respects (see, for example Section 2.7.4). Reading too much significance into the similarities between laws on numeric and Boolean operations can lead you astray.

The essence of algebra is *not* that there are fundamental addition and multiplication operators that appear everywhere. The essential idea is that we can use equations to state axioms on a set of operators, and then use equational reasoning to explore the properties of the resulting system. Some algebraic systems have addition and multiplication operators, and some algebraic systems don't. Boolean algebra doesn't.

Table 2.1 summarises the laws of Boolean Algebra, and we'll discuss them in more detail in the following sections. If we were mainly concerned with the foundations of algebra, our aim would be to take the smallest possible set of equations as axioms, and to derive other ones as theorems. However, we are more concerned here with the practical application of Boolean algebra in computer science, so Table 2.1 gives a richer set of laws that are easier to use for practical calculation than a minimal set of axioms would be.

2.7.2   Operations with Constants

These simple laws describe how $\wedge$ and $\vee$ interact with the Boolean constants True and False.

Table 2.1: Laws of Boolean Algebra

$a \wedge \text{False}$	$=$	False	$\{\wedge \text{ null}\}$
$a \vee \text{True}$	$=$	True	$\{\vee \text{ null}\}$
$a \wedge \text{True}$	$=$	a	$\{\wedge \text{ identity}\}$
$a \vee \text{False}$	$=$	a	$\{\vee \text{ identity}\}$
a	$\to$	$a \vee b$	$\{\text{disjunctive implication}\}$
$a \wedge b$	$\to$	a	$\{\text{conjunctive implication}\}$
$a \wedge a$	$=$	a	$\{\wedge \text{ idempotent}\}$
$a \vee a$	$=$	a	$\{\vee \text{ idempotent}\}$
$a \wedge b$	$=$	$b \wedge a$	$\{\wedge \text{ commutative}\}$
$a \vee b$	$=$	$b \vee a$	$\{\vee \text{ commutative}\}$
$(a \wedge b) \wedge c$	$=$	$a \wedge (b \wedge c)$	$\{\wedge \text{ associative}\}$
$(a \vee b) \vee c$	$=$	$a \vee (b \vee c)$	$\{\vee \text{ associative}\}$
$a \wedge (b \vee c)$	$=$	$(a \wedge b) \vee (a \wedge c)$	$\{\wedge \text{ distributes over } \vee\}$
$a \vee (b \wedge c)$	$=$	$(a \vee b) \wedge (a \vee c)$	$\{\vee \text{ distributes over } \wedge\}$
$\neg(a \wedge b)$	$=$	$\neg a \vee \neg b$	$\{\text{DeMorgan's law}\}$
$\neg(a \vee b)$	$=$	$\neg a \wedge \neg b$	$\{\text{DeMorgan's law}\}$
$\neg\text{True}$	$=$	False	$\{\text{negate True}\}$
$\neg\text{False}$	$=$	True	$\{\text{negate False}\}$
$a \wedge \neg a$	$=$	False	$\{\wedge \text{ complement}\}$
$a \vee \neg a$	$=$	True	$\{\vee \text{ complement}\}$
$\neg(\neg a)$	$=$	a	$\{\text{double negation}\}$
$a \wedge (a \to b)$	$\to$	b	$\{\textit{Modus Ponens}\}$
$(a \to b) \wedge \neg b$	$\to$	$\neg a$	$\{\textit{Modus Tollens}\}$
$(a \vee b) \wedge \neg a$	$\to$	b	$\{\text{disjunctive syllogism}\}$
$(a \to b) \wedge (b \to c)$	$\to$	$a \to c$	$\{\text{implication chain}\}$
$(a \to b) \wedge (c \to d)$	$\to$	$(a \wedge c) \to (b \wedge d)$	$\{\text{implication combination}\}$
$(a \wedge b) \to c$	$=$	$a \to (b \to c)$	$\{\text{Currying}\}$
$a \to b$	$=$	$\neg a \vee b$	$\{\text{implication}\}$
$a \to b$	$=$	$\neg b \to \neg a$	$\{\text{contrapositive}\}$
$(a \to b) \wedge (a \to \neg b)$	$=$	$\neg a$	$\{\text{absurdity}\}$
$a \leftrightarrow b$	$=$	$(a \to b) \wedge (b \to a)$	$\{\text{equivalence}\}$

$$
\begin{array}{rcll}
a \wedge \text{False} & = & \text{False} & \{\wedge \text{ null}\} \\
a \vee \text{True} & = & \text{True} & \{\vee \text{ null}\} \\
a \wedge \text{True} & = & a & \{\wedge \text{ identity}\} \\
a \vee \text{False} & = & a & \{\vee \text{ identity}\}
\end{array}
$$

Often it's possible to simplify Boolean expressions with equational reasoning using the constant laws. If you already know what the final simplified result will be, then the equational reasoning serves as a proof of the equation. Here, for example, is a simplification of the expression $(P \wedge \text{True}) \vee \text{False}$. Alternatively, the following reasoning is a proof of the equation $(P \wedge \text{True}) \vee \text{False} = P$.

$$
\begin{array}{ll}
(P \wedge \text{True}) \vee \text{False} & \\
\quad = P \wedge \text{True} & \{\vee \text{ identity}\} \\
\quad = P & \{\wedge \text{ identity}\}
\end{array}
$$

Note the form of the proof. We are trying to prove an equation, and the proof consists of a chain of equations. The chain begins with the left hand side of the theorem and ends with the right hand side of the theorem. Each step of the chain is justified by one of the laws of Boolean algebra, and the name of the law is written to the right.

Exercise 18. Simplify $(P \wedge \text{False}) \vee (Q \wedge \text{True})$.

Exercise 19. Prove the equation $(P \wedge \text{False}) \wedge \text{True} = \text{False}$.

2.7.3   Basic Properties of $\wedge$ and $\vee$

The following laws describe the basic properties of the $\wedge$ and $\vee$ operators. An *idempotent* property allows you to collapse expressions like $a \wedge a \wedge a$ down to just a. Commutativity means that the order of the operands can be reversed, and associativity means that the grouping of parentheses can be changed without affecting the meaning.

$$
\begin{array}{rcll}
a & \rightarrow & a \vee b & \{\text{disjunctive implication}\} \\
a \wedge b & \rightarrow & a & \{\text{conjunctive implication}\} \\
a \wedge a & = & a & \{\wedge \text{ idempotent}\} \\
a \vee a & = & a & \{\vee \text{ idempotent}\} \\
a \wedge b & = & b \wedge a & \{\wedge \text{ commutative}\} \\
a \vee b & = & b \vee a & \{\vee \text{ commutative}\} \\
(a \wedge b) \wedge c & = & a \wedge (b \wedge c) & \{\wedge \text{ associative}\} \\
(a \vee b) \vee c & = & a \vee (b \vee c) & \{\vee \text{ associative}\}
\end{array}
$$

Commutative operators take two operands, but the order doesn't matter. The commutative properties are often needed to put an expression into a form where you can use another of the identities. Although we don't have a law saying that $\mathsf{False} \wedge a = \mathsf{False}$, the commutativity of $\wedge$ can be applied to rearrange an expression so that the law we do have, $a \wedge \mathsf{False} = \mathsf{False}$, becomes usable. As an example, here is a proof of the equation $(\mathsf{False} \wedge P) \vee Q = Q$:

$(\mathsf{False} \wedge P) \vee Q$
$\quad = (P \wedge \mathsf{False}) \vee Q \qquad\qquad\qquad \{\wedge \text{ commutative}\}$
$\quad = \mathsf{False} \vee Q \qquad\qquad\qquad\quad \{\wedge \text{ null}\}$
$\quad = Q \vee \mathsf{False} \qquad\qquad\qquad\quad \{\vee \text{ commutative}\}$
$\quad = Q \qquad\qquad\qquad\qquad\qquad \{\vee \text{ identity}\}$

An associative operator gives the same result regardless of grouping. For example, ordinary addition is associative, so $2 + (3 + 4) = (2 + 3) + 4$. In a similar way, $\wedge$ and $\vee$ are both associative.

Since the parentheses don't actually affect the result in an expression where an associative operator is repeated, you can safely omit them. For example, we commonly write expressions like $2 + x + y$, without insisting on $2 + (x + y)$ or $(2 + x) + y$. The same thing happens in Boolean algebra: the propositions $P \wedge Q \wedge R$ and $A \vee B \vee C \vee D$ are unambigous because the $\wedge$ and $\vee$ operators are associative, and it makes no difference what order the operations are performed. When you mix different operators, however, parentheses are important: $P \wedge (Q \vee R)$ is not the same thing as $(P \wedge Q) \vee R$.

Exercise 20. Prove $(P \wedge ((Q \vee R) \vee Q)) \wedge S = S \wedge ((R \vee Q) \wedge P)$.

Exercise 21. Prove $P \wedge (Q \wedge (R \wedge S)) = ((P \wedge Q) \wedge R) \wedge S$.

2.7.4 Distributive and DeMorgan's Laws

The laws in this section describe some important properties of expressions that contain both the $\vee$ and $\wedge$ operators.

$a \wedge (b \vee c)$	$=$	$(a \wedge b) \vee (a \wedge c)$	$\{\wedge \text{ distributes over } \vee\}$
$a \vee (b \wedge c)$	$=$	$(a \vee b) \wedge (a \vee c)$	$\{\vee \text{ distributes over } \wedge\}$
$\neg(a \wedge b)$	$=$	$\neg a \vee \neg b$	$\{\text{DeMorgan's law}\}$
$\neg(a \vee b)$	$=$	$\neg a \wedge \neg b$	$\{\text{DeMorgan's law}\}$

The distributive laws are analogous to the way multiplication distributes over addition in elementary algebra: $a \times (b+c) = (a \times b) + (a \times c)$. However, it is *not* the case that addition distributes over multiplication, because $a + (b \times c) \neq (a + b) \times (a + c)$. Here is a significant reason that you should not think of $\vee$ and $\wedge$ as being addition and multiplication.

There is an intuitive reading for both of DeMorgan's laws. The proposition $\neg(a \wedge b)$ says 'a and b aren't *both* true'. An equivalent way to say this is 'either a or b must be false,' which corresponds to $\neg a \vee \neg b$.

Exercise 22. Give an intuitive explanation of the second DeMorgan's law.

2.7.5   Laws on Negation

The following laws state some simple properties of logical negation ($\neg$). We'll
see some more subtle properties in the following section, where negation is
mixed with implication.

$$
\begin{array}{rcll}
\neg\mathsf{True} & = & \mathsf{False} & \{\text{negate True}\} \\
\neg\mathsf{False} & = & \mathsf{True} & \{\text{negate False}\} \\
a \wedge \neg a & = & \mathsf{False} & \{\wedge \text{ complement}\} \\
a \vee \neg a & = & \mathsf{True} & \{\vee \text{ complement}\} \\
\neg(\neg a) & = & a & \{\text{double negation}\}
\end{array}
$$

The following example shows how equational reasoning can be used to simplify $P \wedge \neg(Q \vee P)$:

$$
\begin{array}{ll}
P \wedge \neg(Q \vee P) & \\
\quad = P \wedge (\neg Q \wedge \neg P) & \{\text{DeMorgan's law}\} \\
\quad = P \wedge (\neg P \wedge \neg Q) & \{\wedge \text{ commutative}\} \\
\quad = (P \wedge \neg P) \wedge \neg Q & \{\wedge \text{ associative}\} \\
\quad = \mathsf{False} \wedge \neg Q & \{\wedge \text{ complement}\} \\
\quad = \neg Q \wedge \mathsf{False} & \{\wedge \text{ commutative}\} \\
\quad = \mathsf{False} & \{\wedge \text{ null}\}
\end{array}
$$

2.7.6   Laws on Implication

The laws on implication are frequently useful for solving problems, and they are
also subtle enough to warrant careful study—especially the ones that combine
the $\rightarrow$ and $\neg$ operators. Note that some of these laws are implications, and
others are equations. A good way to understand an implication law is to find
a counterexample demonstrating that it would not be valid as an equation.
For example, the conjunctive implication law says that the conjunction $a \wedge b$
implies a. However, it is *not* valid to write the implication in the other direction:
$a \rightarrow (a \wedge b)$ is False when $a = \mathsf{True}$ and $b = \mathsf{False}$.

$$
\begin{array}{rcll}
a \wedge (a \rightarrow b) & \rightarrow & b & \{\textit{Modus Ponens}\} \\
(a \rightarrow b) \wedge \neg b & \rightarrow & \neg a & \{\textit{Modus Tollens}\} \\
(a \vee b) \wedge \neg a & \rightarrow & b & \{\text{disjunctive syllogism}\} \\
(a \rightarrow b) \wedge (b \rightarrow c) & \rightarrow & a \rightarrow c & \{\text{implication chain}\} \\
(a \rightarrow b) \wedge (c \rightarrow d) & \rightarrow & (a \wedge c) \rightarrow (b \wedge d) & \{\text{implication combination}\} \\
(a \wedge b) \rightarrow c & = & a \rightarrow (b \rightarrow c) & \{\text{Currying}\} \\
a \rightarrow b & = & \neg a \vee b & \{\text{implication}\} \\
a \rightarrow b & = & \neg b \rightarrow \neg a & \{\text{contrapositive}\} \\
(a \rightarrow b) \wedge (a \rightarrow \neg b) & = & \neg a & \{\text{absurdity}\}
\end{array}
$$

Consider the Currying law, which is a logical form of Curried function arguments (which will be covered in Chapter 9). Suppose two conditions a and b are sufficient to ensure that c must be true. The Currying law says, in effect, that there are two equivalent ways to establish that a and b both hold: either we can require that $a \wedge b$ is true, or we can require that an implication on a is satisfied and also an implication on b. If either a or b is false, then $a \wedge b$ will be false, so the implication $(a \wedge b) \rightarrow c$ is vacuous. Furthermore at least one of the implications in $a \rightarrow (b \rightarrow c)$ will also be vacuous.

The second law, $a \rightarrow b = \neg a \vee b$, often provides the easiest way to prove implications in Boolean algebra. Notice that one side of the equation contains the $\rightarrow$ operator and the other doesn't; therefore you can use this equation to introduce a $\rightarrow$ where none was present before. Boolean algebra doesn't have any notion of inference, so you can't prove a proposition containing a $\rightarrow$ with an introduction rule.

The contrapositive law lets you turn around an implication. Suppose you know that $a \rightarrow b$; then if b is false it can't be the case that a is true.

The absurdity law is quite powerful, because it allows us to deduce the value of a just from implications on a. This is worth thinking about: you might expect that an implication $a \rightarrow b$ tells you something about b *if you know* a, but it can't tell you whether a is true. Suppose, however, we know both $a \rightarrow b$ and also $a \rightarrow \neg b$. Then a can't be true because $b \wedge \neg b$ can't be true.

2.7.7 Equivalence

Strictly speaking, we don't need the logical equivalence operator $\leftrightarrow$ at all. The proposition $a \leftrightarrow b$ is simply an abbreviation for $(a \rightarrow b) \wedge (b \rightarrow a)$, as stated by the following equation.

$$a \leftrightarrow b \quad = \quad (a \rightarrow b) \wedge (b \rightarrow a) \quad \{\text{equivalence}\}$$

Logical equivalence is essentially similar to equality, but there is a subtle distinction. The well-formed formula $P \wedge P \leftrightarrow P$ is a single proposition, which happens to have the value True. In contrast, the equation $P \wedge P = P$ is not a proposition. It is an equation, whose left and right hand sides are propositions. The equation is a statement in the metalanguage *about* propositions which are expressed in the object language. You can say that two propositions have the same value in either language; if you're in the metalanguage, talking about propositions, use $=$, but if you're in the object language, trying to write propositions that express properties of other propositions, use $\leftrightarrow$.

2.8 Logic in Computer Science

Logic and computer science are strongly connected subjects. Each one has a major influence on the other. The proof checking software described in Section

2.6 is a typical example of the application of computing to logic. In this section we're concerned with the other direction: the benefits of logic to computing. There are far too many applications of logic to computing to mention them all here. This section describes just a few examples in order to help put the topics in this chapter into perspective. Section 2.10 gives references that will tell you more about these topics.

Formal correctness of software.   Many large programming projects have been devastated by ineradicable bugs. There has been much discussion of the 'Software Crisis': how can we write software that works correctly? One approach to this problem is to use mathematics to specify what the software is supposed to do, and to prove its correctness. There are many different methods for doing this, but ultimately they are all based on formal logical reasoning.

In general it doesn't work very well to take a poorly-written program and try to prove it correct. Even if the program contains no bugs, the sloppy structure will make the logical correctness proof impossibly difficult. However, there has been considerable success in using the formal logical reasoning to help derive the software from a mathematical specification.

The Curry-Howard Isomorphism and type systems.   Many modern programming languages—especially functional languages like Haskell—have powerful and expressive type systems. We need effective methods to help deal with type systems: to help programmers understand them, to help compiler writers implement them, and to help language designers to specify them.

There is a remarkable connection between the inference rules of logic and the typing rules of programming languages; in fact, they are essentially the same! This connection was observed in the 1950s by the logicians Curry and Howard, and has ultimately resulted in great improvements in programming languages.

Linear Logic and access control.   Often in computing we are concerned with controlling the access to some resource. One example arises in functional programming, where array operations can be implemented more efficiently if the compiler can guarantee that the program observes certain constraints on the way the array is accessed.

There is a logical system called Linear Logic which keeps track of where each intermediate result in an inference proof is used. The inference rules of linear logic are careful to discharge every assumption exactly once, and intermediate results must also be discharged. The system is able to express certain kinds of resource utilisation through the inference rules. There is significant current research on the application of linear logic to language design and compilers.

Digital hardware design.   Computers are built out of digital hardware. These circuits are very complex, containing enormous numbers of components.

Discrete mathematics is an essential tool for designing digital circuits. Using the mathematics makes the design process easier, and also makes it possible to prove that a circuit is correct. In Chapter 10 we will return to this application in more detail.

2.9 Metalogic

Metalogic is concerned with stepping outside the language of logic, so that we can make general statements about the properties of the logical system itself. Metalogic enables us to talk *about* logic rather than just saying things *in* logic. Within the logical system, the only elements of our vocabulary are the propositional variables and logical operators; we can say things like $A \wedge B \to A \vee B$ but that's all. The purpose of metalogic is to enable us to think about deeper questions.

At this point we have covered three quite different methods for reasoning about propositions: truth tables, logical inference and Boolean algebra. These methods have completely different styles:

- Truth tables enable us to calculate the values (or *meanings*) of propositions, given the values of their variables. The basic technique is calculation, and it results in a logical value (True or False). The meaning of an expression is called its *semantics*, so calculation with truth tables is a form of semantic reasoning.

- Inference rules enable us to prove theorems. The basic technique involves matching the structures of propositions with the structure of the formulas in the inference rules. The structure of an expression is called its *syntax*, so logical inference is a form of syntactic reasoning.

- Boolean algebra allows the use of equational reasoning to prove the equality of two expressions, or to calculate the values of expressions. It applies the power of algebra to the subject of logic.

Just as propositional logic needs a vocabulary for talking about truth values ($\wedge$, $\vee$ etc.), metalogic needs a vocabulary for talking about logical reasoning itself. There are two fundamental operator symbols in metalogic, $\models$ and $\vdash$, which correspond to semantic and syntactic reasoning respectively. We have already defined these operators. Recall that:

- $P_1, P_2, \ldots, P_n \vdash Q$ means that there is a proof which infers the conclusion Q from the assumptions $P_1, \ldots, P_n$ using the formal inference rules of natural deduction;

- $P_1, P_2, \ldots, P_n \models Q$ means that Q must be True if $P_1, \ldots, P_n$ are all True, but it says nothing about whether we have a proof, or indeed even whether a proof is possible.

The $\models$ and the $\vdash$ operators are describing two different notions of truth, and it's important to know whether they are in fact equivalent. Is it possible to prove a theorem which is false? Is there a true theorem for which no proof exists?

Definition 6. A formal system is *consistent* if the following statement is true for all well-formed formulas a and b:

$$\text{If } a \vdash b \text{ then } a \models b.$$

In other words, the system is consistent if each proposition provable using the inference rules is *actually* true.

Definition 7. A formal system is *complete* if the following statement is true for all well formed formulas a and b:

$$\text{If } a \models b \text{ then } a \vdash b.$$

In other words, the system is complete if the inference rules are powerful enough to prove every proposition which is true.

Fortunately it turns out that propositional logic is both consistent and complete. This means that you can't prove false theorems and if a theorem is true it has a proof.

Theorem 22. Propositional logic is consistent and complete.

You can find the proof of Theorem 22 in some of the books suggested in the next section. It is interesting to note, however, that this is a (metalogical) theorem which is proved mathematically; it isn't just an assumption.

There are many logical systems, and some of them are much richer and more expressive than propositional logic. In the next chapter we will look at a more powerful logical system called *predicate logic*, which is also consistent and complete.

It turns out that any logical system that is powerful enough to express ordinary arithmetic must be either inconsistent or incomplete. This means it's impossible to capture all of mathematics in a safe logical system. This result, the famous *Gödel's Theorem*, has had a profound impact on the philosophy of mathematics, and is also relevant to theoretical computer science.

2.10   Suggestions for Further Reading

Logic plays an increasingly important role in computer science, and there are many books and papers where you can learn more about the connections between these two subjects. There is a regular international conference, *Logic in Computer Science* (LICS), devoted to current research. A number of recommendations for further reading are given below; the complete citations appear in the Bibliography.

- *Forever Undecided: A Puzzle Guide to Gödel*, by Raymond Smullyan [26]. This is a collection of puzzles set on an island of Knights and Knaves. Knights are always truthful, and knaves always lie. The tricky problem is that you can't tell whether someone is a knight or a knave simply by looking at them. A typical problem is to think of a question to ask someone which will enable you to figure out what you want to know, regardless of whether the person you ask happens to be a knight or a knave. Smullyan manages to capture the essence of several metalogical problems, including the famous Incompleteness Theorem of Gödel, and to express deep problems in terms of entertaining and relatively elementary puzzles. This book is a classic of logic and philosophy. Don't miss it!

- *Gödel, Escher, Bach: An Eternal Golden Braid*, by Douglas R. Hofstadter [16] is 'A metaphorical fugue on minds and machines in the spirit of Lewis Carroll.' The themes of the book are drawn from mathematics, art, music, cognitive science and computer science. Another unmissable classic.

- *How To Prove It: A Structured Approach*, by Daniel J. Velleman [31], is a systematic presentation of the standard methods for logical reasoning in carrying out proofs. This is a good source for hints on technique, and it contains lots of examples.

- *Logic for Mathematics and Computer Science*, by Stanley N. Burris [5] is a more advanced coverage of mathematical logic. It gives detailed presentations of some important proof techniques that are useful for automated systems. There is also a good explanation of some interesting historical topics, including syllogisms and Boole's original attempt to apply algebra to logic.

- *A Mathematical Introduction to Logic*, by Herbert B. Enderton [9] gives a standard presentation of logic from a mathematical perspective, including advanced topics such as models, soundness and completeness, and undecidability.

- *Type Theory and Functional Programming*, by Simon Thompson [29] gives a detailed development of the relationship between the inference rules of natural deduction and the rules used to define type systems for programming languages.

- *Logic and Declarative Language*, by Michael Downward [8] covers the relationship between logic and declarative languages, especially logic programming languages like Prolog.

- *Proofs and Types*, by Girard, Lafont and Taylor [12] presents the relationship between inference proofs and type systems at a research level.

2.11   Review Exercises

Exercise 23. Prove $P, Q, R, S \vdash (P \wedge Q) \wedge (R \wedge S)$.

Exercise 24. Prove $P \rightarrow R \vdash P \wedge Q \rightarrow R$.

Exercise 25. Use the inference rules to calculate the value of True $\vee$ True.

Exercise 26. Use the inference rules to calculate the value of False $\rightarrow$ True.
Many people find the truth table definition False $\rightarrow$ True $=$ True to be
highly counterintuitive. Back in Section 2.4, this was just an arbitrary
definition. Now, however, you can see the real reason behind the truth
table definition of $\rightarrow$: it's necessary in order to make the semantic truth
table definition behave consistently with the formal inference rules of
propositional logic.

Exercise 27. Suppose that you were given the following code:

```
logicExpr1 :: Bool -> Bool -> Bool
logicExpr1 a b = a /\ b \/ a <=> a


logicExpr2 :: Bool -> Bool -> Bool
logicExpr2 a b = (a \/ b) /\ b <=> a /\ b
```

Each of these functions specifies a Boolean expression. What are the truth
values of these expressions? How would you write a list comprehension
that can calculate the values for you to check your work?

Exercise 28. Work out the values of these expressions, then check with a list
comprehension:

```
logicExpr3 :: Bool -> Bool -> Bool -> Bool
logicExpr3 a b c
  = (a /\ b) \/ (a /\ c) ==> a \/ b


logicExpr4 :: Bool -> Bool -> Bool -> Bool
logicExpr4 a b c
  = (a /\ (b \/ c)) \/ (a \/ c) ==> a \/ c
```

Exercise 29. Suppose that you are told to calculate the truth value of the
following expression:

```
(a /\ b \/ c /\ d) /\
(e \/ f /\ (g /\ h))
   <=>
(i ==> j /\ (k ==> i \/ b))
```

Would it be better to use a truth table or simplification using laws?

Exercise 30. Define a data type that represents logical expressions.

Exercise 31. Using the Logic data type given in the previous question, define a function `distribute` that implements the distributive law, and a function `deMorgan` that implements DeMorgan's law. Why doesn't `distribute` have four lines, instead of two?

Exercise 32. A proof of an implication requires that the left-hand side, called the *premise*, be assumed. If it is possible to use laws to derive the right hand side, called the *consequent*, then the derivation is the proof. Prove that

$$C \wedge A \wedge B \vee C \rightarrow C \wedge (C \vee (A \wedge B))$$

is a tautology.

Exercise 33. Prove that

$$C \vee A \wedge (B \vee C) \rightarrow ((C \vee A) \wedge C) \vee A \wedge B$$

is a tautology.

Chapter 3

Predicate Logic

It is frequently necessary to reason logically about statements of the form *everything has the property p* or *something has the property p*. One of the oldest and most famous pieces of logical reasoning, which was known to the ancient Greeks, is an example:

> All men are mortal. Socrates is a man. Therefore Socrates is mortal.

In general, propositional logic is not expressive enough to support such reasoning. We could define a proposition P to mean 'all men are mortal', but P is an atomic symbol—it has no internal structure—so we cannot do any formal reasoning that makes use of the meaning of 'all'.

Predicate logic, also called first order logic, is an extension to propositional logic which adds two *quantifiers* that allow statements like the examples above to be expressed. Everything in propositional logic is also in predicate logic: all the definitions, inference rules, theorems, algebraic laws, etc., still hold.

3.1 The Language of Predicate Logic

The formal language of predicate logic consists of propositional logic, augmented with variables, predicates and quantifiers.

3.1.1 Predicates

A *predicate* is a statement that an object x has a certain property. Such statements may be either true or false. For example, the statement '$x > 5$' is a predicate, and its truth depends on the value of x. A predicate can be extended to several variables; for example, '$x > y$' is a predicate about x and y.

The conditional expressions used to control execution of computer programs are predicates. For example, the Haskell expression if x<0 then -x else x uses the predicate x<0 to make a decision.

In predicate logic it is traditional to write predicates concisely in the form $F(x)$, where F is the predicate and x is the variable it is applied to. A predicate containing two variables could be written $G(x, y)$. A predicate is essentially a function that returns a Boolean result. Often in this book we will use Haskell notation for function applications, which does not require parentheses: for example $f\ x$ is the application of f to x. For predicate logic, we will use the traditional notation with parentheses, $F(x)$.

Definition 8. Any term in the form $F(x)$, where F is a predicate name and x is a variable name, is a well-formed formula. Similarly, $F(x_1, x_2, \ldots, x_k)$ is a well-formed formula; this is a predicate containing k variables.

When predicate logic is used to solve a reasoning problem, the first step is to translate from English (or mathematics) into the formal language of predicate logic. This means the predicates are defined; for example we might define:

$$\begin{aligned} F(x) &\equiv x > 0 \\ G(x, y) &\equiv x > y \end{aligned}$$

The *universe of discourse*, often simply called the *universe* or abbreviated U, is the set of possible values that the variables can have. Usually the universe is specified just once, at the beginning of a piece of logical reasoning, but this specification cannot be omitted. For example, consider the statement 'For every x there exists a y such that $x = 2 \times y$'. If the universe is the set of even integers, or the set of real numbers, then the statement is true. However, if the universe is the set of natural numbers then the statement is false (let x be any odd number). If the universe doesn't contain numbers at all, then the statement is not true or false; it is meaningless.

Several notational conventions are very common in predicate logic, although some authors do not follow them. These standard notations will, however, be used in this book. The universe is called U, and its constants are written as lower case letters, typically c and p (to suggest a constant value, or a particular value). Variables are also lower case letters, typically x, y, z. Predicates are upper case letters F, G, H, ... For example, $F(x)$ is a valid expression in the language of predicate logic, and its intuitive meaning is 'the variable x has the property F'. Generic expressions are written with a lower case predicate; for example $f(x)$ could stand for *any* predicate f applied to a variable x.

3.1.2 Quantifiers

There are two *quantifiers* in predicate logic; these are the special symbols $\forall$ and $\exists$.

Definition 9. If $F(x)$ is a well-formed formula containing the variable x, then $\forall x.\ F(x)$ is a well-formed formula called a *universal quantification*. This is a statement that *everything* in the universe has a certain property: 'For all x in the universe, the predicate $F(x)$ holds'. An alternative reading is 'Every x has the property F'.

Universal quantifications are often used to state required properties. For example, if you want to say formally that a computer program will give the correct output for all inputs, you would use $\forall$. The upside-down A symbol is intended to remind you of *All*.

Example 1. Let U be the set of even numbers. Let $E(x)$ mean x is even. Then $\forall x.\ E(x)$ is a well-formed formula, and its value is true.

Example 2. Let U be the set of natural numbers. Let $E(x)$ mean x is even. Then $\forall x.\ E(x)$ is a well-formed formula, and its value is false.

Definition 10. If $F(x)$ is a well-formed formula containing the variable x, then $\exists x.\ F(x)$ is a well-formed formula called an *existential quantification*. This is a statement that *something* in the universe has a certain property: 'There exists an x in the universe for which the predicate $F(x)$ holds'. An alternative reading is 'Some x has the property F'.

Existential quantifications are used to state properties that must occur at least once. For example, we might want to state that a database contains a record for a certain person; this would be done with $\exists$. The backwards E symbol is reminiscent of *Exists*.

Example 3. Let U be the set of natural numbers. Let $F(x, y)$ be defined as $2 \times x = y$. Then $\exists x.F(x, 6)$ is a well-formed formula; it says that there is a natural number x which gives 6 when doubled; 3 satisfies the predicate, so the formula is true. However, $\exists x.\ F(x, 7)$ is false.

Quantified expressions can be nested. Let the universe be the set of integers, and define $F(x) = \exists y.\ x < y$; thus $F(x)$ means 'There is some number larger than x'. Now we can say that every integer has this property by stating $\forall x.F(x)$. An equivalent way to write this is $\forall x.\ (\exists y.\ x < y)$. The parentheses are not required, since there is no ambiguity in writing $\forall x.\ \exists y.\ x < y$. All the quantified variables must be members of the universe.

In the example above, both x and y are integers. However, it is often useful to have several variables that are members of *different* sets. For example, suppose we are reasoning about people who live in cities, and want to make statements like 'There is at least one person living in every city'. It is natural to define $L(x, y)$ to mean 'The person x lives in the city y', and the expression $\forall x.\exists y.L(y, x)$ then means 'Every city has somebody living in it'. But what is the universe?

The way to handle this problem is to define a separate set of possible values for each variable. For example, let $C = \{\text{London}, \text{Paris}, \text{Los Angeles}, \text{München}\}$ be the set of cities, and let $P = \{\text{Smith}, \text{Jones}, \cdots\}$ be the set of persons. Now we can let the universe contain *all* the possible variable values: $U =$

$C \cup P$. Quantified expressions need to restrict each variable to the set of relevant values, since it is no longer intended that a variable x could be *any* element of U. This is expressed by writing $\forall x \in S.\ F(x)$ or $\exists x \in S.\ F(x)$, which say that x must be an element of S (and therefore also a member of the universe). Now the statement 'There is at least one person living in every city' is written

$$\forall x \in C.\ \exists y \in P.\ L(y, x).$$

Universal quantification over an empty set is vacuously true, and existential quantification over an empty set is vacuously false. Often we require that the universe is non-empty, so quantifications over the universe are not automatically true or false.

3.1.3 Expanding Quantified Expressions

If the universe is finite (or if the variables are restricted to a finite set), expressions with quantifiers can be interpreted as ordinary terms in propositional logic. Suppose $U = \{c_1, c_2, \dots, c_n\}$, where the size of the universe is n. Then quantified expressions can be expanded as follows:

$$\forall x.\ F(x) \quad = \quad F(c_1) \wedge F(c_2) \wedge \cdots \wedge F(c_n) \tag{3.1}$$
$$\exists x.\ F(x) \quad = \quad F(c_1) \vee F(c_2) \vee \cdots \vee F(c_n) \tag{3.2}$$

With a finite universe, therefore, the quantifiers are just syntactic abbreviations. With a small universe it is perfectly feasible to reason directly with the expanded expressions. In many computing applications the universe is finite but may contain millions of elements; in this case the quantifiers are needed to make logical reasoning practical, although they are not needed in principle.

If the variables are not restricted to a finite set, it is impossible even in principle to expand a quantified expression. It may be intuitively clear to write $F(c_1) \wedge F(c_2) \wedge F(c_3) \wedge \cdots$, but this is not a well-formed formula. Every well-formed formula has a finite size, although there is no bound on how large a formula may be. This means that in the presence of an infinite universe, quantifiers make the language of predicate logic more expressive than propositional logic.

The expansion formulas, Equations 3.1 and 3.2, are useful for computing with predicates.

Example 4. Let the universe $U = \{1, 2, 3\}$, and define the predicates *even* and odd as follows:

$$even\ x \quad \equiv \quad (x \bmod 2 = 0)$$
$$odd\ x \quad \equiv \quad (x \bmod 2 = 1)$$

Two quantified expressions will be expanded and evaluated using these definitions:

$\forall x.\ \big(even\ x \to \neg(odd\ x)\big)$
$= (even\ 1 \to \neg(odd\ 1)) \land (even\ 2 \to \neg(odd\ 2)) \land (even\ 3 \to \neg(odd\ 3))$
$= (\mathsf{False} \to \neg\mathsf{True}) \land (\mathsf{True} \to \neg\mathsf{False}) \land (\mathsf{False} \to \neg\mathsf{True})$
$= \mathsf{True} \land \mathsf{True} \land \mathsf{True}$
$= \mathsf{True}$
$\exists x.\ (even\ x \land odd\ x)$
$= (even\ 1 \land odd\ 1) \lor (even\ 2 \land odd\ 2) \lor (even\ 3 \land odd\ 3)$
$= (\mathsf{False} \land \mathsf{True}) \lor (\mathsf{True} \land \mathsf{False}) \lor (\mathsf{False} \land \mathsf{True})$
$= \mathsf{False} \lor \mathsf{False} \lor \mathsf{False}$
$= \mathsf{False}$

Example 5. Let $S = \{0, 2, 4, 6\}$ and $R = \{0, 1, 2, 3\}$. Then we can state that every element of S is twice some element of R as follows:

$$\forall x \in S.\ \exists y \in R.\ x = 2 \times y$$

This can be expanded into a quantifier-free expression in two steps. The first step is to expand the *outer* quantifier:

$$(\exists y \in R.\ 0 = 2 \times y)$$
$$\land\ (\exists y \in R.\ 2 = 2 \times y)$$
$$\land\ (\exists y \in R.\ 4 = 2 \times y)$$
$$\land\ (\exists y \in R.\ 6 = 2 \times y)$$

The second step is to expand all four of the remaining quantifiers:

$$\big((0 = 2 \times 0) \lor (0 = 2 \times 1) \lor (0 = 2 \times 2) \lor (0 = 2 \times 3)\big)$$
$$\land\ \big((2 = 2 \times 0) \lor (2 = 2 \times 1) \lor (2 = 2 \times 2) \lor (2 = 2 \times 3)\big)$$
$$\land\ \big((4 = 2 \times 0) \lor (4 = 2 \times 1) \lor (4 = 2 \times 2) \lor (4 = 2 \times 3)\big)$$
$$\land\ \big((6 = 2 \times 0) \lor (6 = 2 \times 1) \lor (6 = 2 \times 2) \lor (6 = 2 \times 3)\big)$$

Two short cuts have been taken in the notation here. (1) Since every quantified variable is restricted to a set, the universe was not stated explicitly; however we can define $U = S \cup R$. (2) Instead of defining $F(x, y)$ to mean $x = 2 \times y$ and writing $\forall x \in S.\ \exists y \in R.\ F(x, y)$, we simply wrote the expression $x = 2 \times y$ inside the expression. Both short cuts are frequently used in practice.

Exercise 1. Let the universe $U = \{1, 2, 3\}$. Expand the following expressions into propositional term (i.e. remove the quantifiers):

(a) $\forall x.\ F(x)$

(b) $\exists x.\ F(x)$

(c) $\exists x.\ \forall y.\ G(x, y)$

Exercise 2. Let the universe be the set of integers. Expand the following expression: $\forall x \in \{1, 2, 3, 4\}.\ \exists y \in \{5, 6\}.\ F(x, y)$

3.1.4   The Scope of Variable Bindings

Quantifiers *bind* variables by assigning them values from a universe. A dangling expression without explicit quantification, such as $x+2$, has no explicit variable binding. If such an expression appears, x is assumed implicitly to be an element of the universe, and the author should have told you explicitly somewhere what the universe is.

The extent of a variable binding is called its *scope*. For example, the scope of x in the expression $\exists x.\ F(x)$ is the subexpression $F(x)$. For $\forall x \in S.\ \exists y \in R.\ F(x,y)$, the scope of x is $\exists y \in R.\ F(x,y)$, and the scope of y is $F(x,y)$.

It is good practice to use parentheses to make expressions clear and readable. The expression

$$\forall x.\, p(x) \ \lor \ q(x)$$

is not clear: it can be read in two different ways. It could mean either

$$\forall x.\, (p(x) \ \lor \ q(x))$$

or

$$(\forall x.\, p(x)) \ \lor \ q(x).$$

It is probably best to use parentheses in case of doubt, but there is a convention which resolves unclear expressions: the quantifier extends over the smallest subexpression possible unless parentheses indicate otherwise. In other words, the scope of a variable binding is the smallest possible. So, in the assertion given above, the variable x in $q(x)$ is not bound by the $\forall$, so it must have been bound at some outer level (i.e. this expression has to be embedded inside a bigger one).

Often the same quantifier is used several times in a row to define several variables:

$$\forall x.\ \forall y.\ F(x,y)$$

It is common to write this in an abbreviated form, with just one use of the $\forall$ operator followed by several variables separated by commas. For example, the previous expression would be abbreviated as follows:

$$\forall x, y.\ F(x,y)$$

This abbreviation may be used for any number of variables, and it can also be used if the variables are restricted to be in a set, as long as they all have the same restriction. For example, the abbreviated expression

$$\forall x, y, z \in S.\ F(x,y,z)$$

is equivalent to the full expression

$$\forall x \in S.\ \forall y \in S.\ \forall z \in S.\ F(x,y,z).$$

3.1.5 Translating Between English and Logic

Sometimes it is straightforward to translate an English statement into logic. If an English statement has no internal structure that is relevant to the reasoning, it can be represented by an ordinary propositional variable:

$$A \equiv \text{Elephants are big.}$$
$$B \equiv \text{Cats are furry.}$$
$$C \equiv \text{Cats are good pets.}$$

An English statement built up with words like *and, or, not, therefore* and so on, where the meaning corresponds to the logical operators, can be represented by a propositional expression.

$$\neg A \equiv \text{Elephants are small.}$$
$$A \wedge B \equiv \text{Elephants are big and cats are furry.}$$
$$B \rightarrow C \equiv \text{If cats are furry then they make good pets.}$$

Notice in the examples above that no use has been made of the internal structure of the English statements. (The sentence 'elephants are small' may appear to violate this, but it could just as easily have been written 'it is untrue that elephants are big', which corresponds exactly to $\neg A$.)

When general statements are made about classes of objects, then predicates and quantifiers are needed in order to draw conclusions. For example, suppose we try these definitions in propositional logic, without using predicates:

$$A \equiv \text{Small animals are good pets.}$$
$$C \equiv \text{Cats are animals.}$$
$$S \equiv \text{Cats are small.}$$

In ordinary conversation, it would be natural to conclude that cats are good pets, but this cannot be concluded with propositional logic. All we have are three propositions: A, C and S are known, but nothing else, and the only conclusions that can be drawn are uninteresting ones like $A \wedge C$, $S \vee A$, and the like. The substantive conclusion, that cats are good pets, requires reasoning about the internal structure of the English statements. The solution is to use predicates to give a more refined translation of the sentences:

$$A(x) \equiv x \text{ is an animal.}$$
$$C(x) \equiv x \text{ is a cat.}$$
$$S(x) \equiv x \text{ is small.}$$
$$GP(x) \equiv x \text{ is a good pet.}$$

Now a much richer kind of English sentence can be translated into predicate logic:

$$\forall x.\ C(x) \rightarrow A(x) \equiv \text{Cats are animals.}$$
$$\forall x.\ C(x) \rightarrow S(x) \equiv \text{Cats are small.}$$
$$\forall x.\ C(x) \rightarrow S(x) \wedge A(x) \equiv \text{Cats are small animals.}$$
$$\forall x.\ S(x) \wedge A(x) \rightarrow GP(x) \equiv \text{Small animals are good pets.}$$

It is generally straightforward to translate from formal predicate logic into English, since you can just turn each logical operator directly into an English word or phrase. For example,

$$\forall x.\ S(x) \wedge A(x) \rightarrow GP(x)$$

could be translated into English literally:

> (1) For every thing, if that thing is small and that thing is an animal, then that thing is a good pet.

This is graceless English, but at least it's comprehensible and correct. The style can be improved:

> (2) Everything which is small and which is an animal is a good pet.

Even better would be:

> (3) Small animals make good pets.

Such stylistic improvements in the English are optional. It is important to be sure that the effort to improve the literary style doesn't affect the meaning, but this is a question of proper usage of natural language, not of formal logic.

It is sometimes trickier to translate from English into formal logic, precisely because the English usually does not correspond obviously to the logical quantifiers and operators. Sentence (1) above can be translated straightforwardly into logic, sentence (3) is harder. The difficulty is not really in the logic; it is in figuring out exactly what the English sentence says.

Often the real difficulty in translating English into logic is in figuring out what the English says, or what the speaker meant to say. For example, many people make statements like 'All people are not rich'. What this statement *actually says* is

$$\forall x. \neg R(x),$$

where the universe is the set of people and $R(x)$ means 'x is rich'. What is usually meant, however, by such a statement is

$$\neg \forall x.\ R(x),$$

which is quite different from the real meaning. This intended meaning is equivalent to

$$\exists x.\ \neg R(x).$$

Such problems of ambiguity or incorrect grammar in English cannot be solved mathematically, but they do illustrate one of the benefits of mathematics: simply translating a problem from English into formal logic may expose confusion or misunderstanding.

Example 6. Consider the translation of the sentence 'Some birds can fly' into logic. Let the universe be a set that contains all birds (it is all right if it contains other things too, such as frogs and other animals). Let $B(x)$ mean 'x is a bird' and $F(x)$ mean 'x can fly'. Then 'Some birds can fly' is translated as

$$\exists x.\ B(x) \wedge F(x)$$

Warning! A common pitfall is to translate 'Some birds can fly' as

$$\exists x.\ B(x) \rightarrow F(x) \qquad \textit{Wrong translation!}$$

To see why this is wrong, let p be a frog that somehow got into the universe. Now $B(p)$ is false, so $B(p) \rightarrow F(p)$ is true (remember False $\rightarrow$ False = True). This is just saying 'If that frog were a bird then it would be able to fly', which is true; it doesn't mean the frog actually is a bird, or that it actually can fly. However, we have now found a value of x—namely the frog p—for which $B(x) \rightarrow F(x)$ is true, and that is enough to satisfy $\exists x.\ B(x) \rightarrow F(x)$, even if all the birds in the universe happen to be chickens (which cannot fly).

Exercise 3. Express the following statements formally, using the universe of natural numbers, and the predicates $E(x) \equiv x$ is even and $O(x) \equiv x$ is odd.

- There is an even number.

- Every number is either even or odd.

- No number is both even and odd.

- The sum of two odd numbers is even.

- The sum of an odd number and an even number is odd.

Exercise 4. Let the universe be the set of all animals, and define the following predicates:

$$
\begin{array}{rcl}
B(x) & \equiv & x \text{ is a bird.} \\
D(x) & \equiv & x \text{ is a dove.} \\
C(x) & \equiv & x \text{ is a chicken.} \\
P(x) & \equiv & x \text{ is a pig.} \\
F(x) & \equiv & x \text{ can fly.} \\
W(x) & \equiv & x \text{ has wings.} \\
M(x,y) & \equiv & x \text{ has more feathers than } y \text{ does.}
\end{array}
$$

Translate the following sentences into logic. There are generally several correct answers. Some of the English sentences are fairly close to logic, while others require more interpretation before they can be rendered in logic.

- Chickens are birds.

- Some doves can fly.

- Pigs are not birds.
- Some birds can fly, and some can't.
- An animal needs wings in order to fly.
- If a chicken can fly, then pigs have wings.
- Chickens have more feathers than pigs do.
- An animal with more feathers than any chicken can fly.

Exercise 5. Translate the following into English.

- $\forall x. (\exists y.$ wantsToDanceWith $(x, y))$
- $\exists x. (\forall y.$ wantsToPhone $(y, x))$
- $\exists x. ($tired $(x) \land \forall y.$ helpsMoveHouse $(x, y))$

3.2   Computing with Quantifiers

As long as the universe is finite, a computer is useful for evaluating logical expressions with quantifiers. This provides a good way to check your understanding of expressions in predicate logic. Even more importantly, many software applications are expressed in terms of finite sets of data that are manipulated using predicate logic expressions.

The software tools file provides several Haskell functions that are helpful for computing with predicate logic, and this section explains how to use them. To keep things simple, we assume that the universe is a set of numbers represented as a list. In programming terminology, a *predicate* is a function that returns a Boolean value; this is the same meaning that predicate has in logic.

The function `forall` takes a list of numbers forming the universe and a predicate, applies the predicate to each value in the list, and returns the conjunction of the results:

```
forall :: [Int] -> (Int -> Bool) -> Bool
```

For example, `forall [1,2] (>5)` means $\forall x. \ x > 5$ where the universe $U = \{1, 2\}$. The implementation of `forall` simply expands the quantified expression, using Equation 3.1, and then evaluates it. The expansion uses the Haskell function `and ::   [Bool] -> Bool`; thus $F(c_1) \land F(c_2) \land F(c_3)$ would be expressed in Haskell as `and [f c1, f c2, f c3]`.

Exercise 6. Write the predicate logic expressions corresponding to the following Haskell expressions. Then decide whether the value is `True` or `False`, and evaluate using the computer. Note that `(== 2)` is a function that takes a number and compares it with 2, while `(< 4)` is a function that takes a number and returns `True` if it is less than 4.

```
forall [1,2,3] (== 2)
forall [1,2,3] (< 4)
```

Like `forall`, the function `exists` applies its second argument to all of the elements in its first argument:

```
exists :: [Int] -> (Int -> Bool) -> Bool
```

Exercise 7. Again, rewrite the following in predicate logic, work out the values by hand and evaluate on the computer:

```
exists [0,1,2] (== 2)
exists [1,2,3] (> 5)
```

The functions `exists` and `forall` can be nested in the same way as quantifiers can be nested in predicate logic. It's convenient to make inner calls into separate functions.

Example 7. $\forall x \in \{1,2\}. (\exists y \in \{1,2\}. x = y)$ has an inner assertion that can be implemented as follows:

```
inner_fun :: Int -> Int -> Bool
inner_fun x = exists [1,2] (== x)
```

Now consider the evaluation of:

```
forall [1,2] inner_fun
```

The evaluation can be calculated step by step. The function and takes a list of Boolean values and combines them all using the $\wedge$ operation:

```
forall [1,2] inner_fun
= and [inner_fun 1, inner_fun 2]
= and [exists [1,2] (== 1),
       exists [1,2] (== 2)]
= and [or [1==1, 2==1],
       or [1==2, 2==2]]
= and [True, True]
= True
```

Example 8. Define:

```
inner_fun x = exists [1,2,3] (== x+2)
exists [1,2,3] inner_fun
```

Here is the evaluation:

```
exists [1,2,3] inner_fun
= or [inner_fun 1, inner_fun 2, inner_fun 3]
= or [exists [1,2,3] (== 1+2),
      exists [1,2,3] (== 2+2),
      exists [1,2,3] (== 3+2)]
```

Figure 3.1: Inference Rules of Predicate Logic

```
= or [or [1 == 1+2, 2 == 1+2, 3 =÷ 1+2],
      or [1 == 2+2, 2 == 2+2, 3 == 2+2],
      or [1 == 3+2, 2 == 3+2, 3 == 3+2]]
= or [or [False, False, True],
      or [False, False, False],
      or [False, False, False]]
= or [True, False, False]
= True
```

An important distinction between mathematical quantification and the Haskell functions `exists` and `forall` is that quantification is defined over both finite and infinite universes, whereas these Haskell functions do not always terminate when applied to infinite universes.

Exercise 8. Define the predicate $p\ x\ y$ to mean $x = y + 1$, and let the universe be $\{1, 2\}$. Calculate the value of each of the following expressions, and then check your solution using Haskell.

    (a) $\forall x.\ (\exists y.\ p(x, y))$

    (b) $\exists x, y.\ p(x, y)$

    (c) $\exists x.\ (\forall y.\ p(x, y))$

    (d) $\forall x, y.\ p(x, y)$

3.3 Logical Inference with Predicates

The inference rules for propositional logic can be extended to handle predicate logic as well. Four additional rules are required (Figure 3.1): an introduction rule and an elimination rule for both of the quantifiers $\forall$ and $\exists$.

    A good way to understand the inference rules of predicate logic is to view them as generalisations of the corresponding rules of propositional logic. For example, there is a similarity between inferring $F(P) \wedge F(Q)$ in propositional logic, and inferring $\forall x.F(x)$ in predicate logic. If the universe is finite, then the predicate logic is not, in principle, even necessary. We could express $\forall x.F(x)$ by $F(p_1) \wedge F(p_2) \wedge \ldots \wedge F(p_n)$, where n is the size of the universe. If the universe is infinite, however, then the inference rules of predicate logic allow deductions that would be impossible using just the propositional logic rules. We will always assume that the universe is non-empty.

3.3.1 Universal Introduction $\{\forall I\}$

A standard proof technique, which is used frequently in mathematics and computer science, is to state and prove a property about an arbitrary value x, which is an element of the universe, and then to interpret this as a statement about *all* elements of the universe. A typical example is the following simple theorem about Haskell lists, which says that there are two equivalent methods for attaching a singleton x in front of a list xs:

Theorem 23. Let $x :: a$ and $xs :: [a]$. Then $x : xs = [x] \mathbin{+\!\!+} xs$.

It is important to realise that this theorem is stating a property about *arbitrary* x and xs. It really means that the property holds for *all* values of the variables, and this could be stated more formally with an explicit quantifier:

Theorem 24. $\forall x :: a.\ \forall xs :: [a].\ x : xs = [x] \mathbin{+\!\!+} xs$

These two theorems have exactly the same meaning; the only difference is the style[1] in which they are expressed: the first is a little more like English, and the second is a little more formal. Both styles are common. For a theorem in the first style, the use of *arbitrary* variables means an implicit $\forall$ is meant. Now, consider the proof of this theorem; the following proof could be used for either the formal or the informal statement of the theorem:

Proof.

$$
\begin{aligned}
&[x] \mathbin{+\!\!+} xs \\
&= (x : [\,]) \mathbin{+\!\!+} xs \qquad &&\text{def. of notation} \\
&= x : ([\,] \mathbin{+\!\!+} xs) \qquad &&(\mathbin{+\!\!+}).2 \\
&= x : xs \qquad &&(\mathbin{+\!\!+}).1
\end{aligned}
$$

$\square$

Again, there is a significant point about this proof: it consists of formal reasoning about one value x and one value xs, but these values are *arbitrary*, and the conclusion we reach at the end of the proof is that the theorem is true for *all* values of the variables. In other words, if we can prove a theorem for an arbitrary variable, then we infer that the theorem is true for *all* possible values of that variable.

---

[1] For a discussion about good style in mathematics, see the pointer to *Mathematical Writing* in Section 3.5.

These ideas are expressed formally by the following inference rule, which says that if the expression a (which may contain a variable x) can be proved for *arbitrary* x, then we may infer the proposition $\forall x.\ a$. Since this rule specifies what we need to know in order to infer an expression containing $\forall$, its name is $\{\forall I\}$.

$$\frac{F(x) \quad \{x \text{ arbitrary}\}}{\forall x.F(x)}\{\forall I\}$$

To clarify exactly what this rule means, we will compare two examples: one where it can be used, and one where it cannot. Let the universe be the set of natural numbers N, and let $E(x)$ be the proposition 'x is even'. First, consider the following theorem:

Theorem 25. $\vdash \forall x.\ E(x) \rightarrow \big(E(x) \vee \neg E(x)\big)$

The proof uses the $\forall$ introduction rule. The important point is that this inference does not depend on the particular value of p; thus the value of p is arbitrary, and the $\{\forall I\}$ rule allows us to infer $\forall x.F(x) \vee \neg F(x)$.

Proof.

$$\frac{\dfrac{\boxed{E(p)}}{E(p) \vee \neg E(p)}\{\vee I_L\}}{\dfrac{E(p) \rightarrow E(p) \vee \neg E(p)}{\forall x.\ E(x) \rightarrow E(x) \vee \neg E(x)}\{\forall I\}}\{\rightarrow I\}$$

$\square$

Now consider the following *incorrect* proof, which purports to show that all natural numbers are even:

$$\frac{\boxed{E(2)}}{\forall x.E(x)}\{\forall I\} \quad \textbf{\textit{Wrong!}}$$

The theorem $E(2)$ is established for a particular value, 2. However, 2 is not *arbitrary*: the proof that 2 is even relies on its value, and we could not substitute 3 without invalidating it. Since we have not proved $E(x)$ for an arbitrary value, the requirements of the $\{\forall I\}$ inference rule are not satisfied, and we cannot conclude $\forall x.E(x)$.

3.3.2 Universal Elimination $\{\forall E\}$

The universal elimination rule says that if you have established $\forall x.F(x)$, and p is a particular element of the universe, then you can infer $F(p)$.

The following theorem allows you to apply a universal implication, in the form $\forall x.F(x) \rightarrow G(x)$, to a particular proposition $F(p)$, and its proof illustrates the $\{\forall E\}$ inference rule.

Theorem 26. $F(p), \forall x.F(x) \rightarrow G(x) \vdash G(p)$

Proof.

$$
\cfrac{F(p) \quad \cfrac{\forall x.F(x) \rightarrow G(x)}{F(p) \rightarrow G(p)}\{\forall E\}}{G(p)}\{\rightarrow E\}
$$

$\square$

In the previous chapter the implication chain theorem was proved; this says that from $a \rightarrow b$ and $b \rightarrow c$ you can infer $a \rightarrow c$. The $\{\forall I\}$ inference rule can be used to prove the corresponding theorem on universal implications: from $\forall x.F(x) \rightarrow G(x)$ and $\forall x.G(x) \rightarrow H(x)$, you can infer $\forall x.F(x) \rightarrow H(x)$. However, in order to use $\{\forall I\}$ we have to establish first, for an arbitrary p in the universe, that $F(p) \rightarrow H(p)$, and the proof of that proposition requires using the $\{\forall E\}$ rule twice to prove the particular propositions $F(p) \rightarrow G(p)$ and $G(p) \rightarrow H(p)$.

Theorem 27. $\forall x.F(x) \rightarrow G(x), \forall x.G(x) \rightarrow H(x) \vdash \forall x.F(x) \rightarrow H(x)$

Proof.

$$
\cfrac{\cfrac{\cfrac{\forall x.F(x) \rightarrow G(x)}{F(p) \rightarrow G(p)}\{\forall E\} \quad \cfrac{\forall x.G(x) \rightarrow H(x)}{G(p) \rightarrow H(p)}\{\forall E\}}{F(p) \rightarrow H(p)}\text{Th. Imp Chain}}{\forall x.F(x) \rightarrow H(x)}\{\forall I\}
$$

$\square$

The following theorem says that you can change the order in which the variables are bound in $\forall x. \forall y. F(x,y)$. This theorem is simple, but extremely important.

Theorem 28. $\forall x. \forall y. F(x,y) \vdash \forall y. \forall x. F(x,y)$

Proof.

$$\dfrac{\dfrac{\dfrac{\forall x.\,\forall y.\,F(x,y)}{\forall y.\,F(p,y)}\{\forall E\}}{F(p,q)}\{\forall E\}}{\dfrac{\forall x.\,F(x,q)}{\forall y.\,\forall x.\,F(x,y)}\{\forall I\}}\{\forall I\}$$

□

This theorem says that if, for all x, a proposition P implies $f(x)$, then P implies $\forall x.f(x)$. This allows you to pull a proposition P, which does not use x, out of an implication bound by $\forall$.

Theorem 29. $\forall x.\ P \rightarrow f(x) \ \vdash\ P \rightarrow \forall x.\ f(x)$

Proof.

$$\dfrac{\dfrac{\dfrac{\boxed{P}\quad\dfrac{\forall x.\ P \rightarrow f(x)}{P \rightarrow f(c)}\{\forall E\}\ (c\text{ arbitrary})}{f(c)}\{\rightarrow E\}}{\forall x.\ f(x)}\{\forall I\}}{P \rightarrow \forall x.\ f(x)}\{\rightarrow I\}$$

□

Exercise 9. Prove $\exists x.\ \exists y.\ F(x,y) \ \vdash\ \exists y.\ \exists x.\ F(x,y)$.

Exercise 10. Prove $\forall x.F(x), \forall x.F(x) \rightarrow G(x) \vdash \forall x.G(x)$.

3.3.3   Existential Introduction $\{\exists I\}$

The $\{\exists I\}$ rule says that if $f(p)$ has been established for a particular p, then you can infer $\exists x.f(x)$.

$$\boxed{\ \dfrac{f(p)}{\exists x.f(x)}\{\exists I\}\ }$$

The following theorem says that if $F(x)$ holds for all elements of the universe, then it must hold for one of them. Recall that we require the universe of discourse to be non-empty; otherwise this theorem would not hold.

Theorem 30. $\forall x.F(x) \vdash \exists x.F(x)$

Proof.

$$\frac{\dfrac{\forall x.F(x)}{F(p)}\{\forall E\}}{\exists x.F(x)}\{\exists I\}$$

□

3.3.4 Existential Elimination $\{\exists E\}$

Recall the $\{\vee E\}$ inference rule of propositional logic; this says that if you know $a \vee b$, and also that c follows from a and c follows from b, then you can infer c.

If the universe is finite, then $\exists x.F(x)$ can be expressed in the form $F(p_1) \vee \ldots \vee F(p_n)$, where the universe is $\{p_1, \ldots, p_n\}$. We could extend the $\{\vee E\}$ rule so that if we know that $F(p_i)$ holds for some i, and furthermore that A must hold if $F(x)$ holds for *arbitrary* x, then A can be inferred.

The existential elimination rule $\{\exists E\}$ captures this idea, and it provides a much more convenient tool for reasoning than repeated applications of $\{\vee E\}$. Its fundamental importance, however, is that $\{\exists E\}$ may also be used if the universe is infinite. This means it is more powerful than $\{\vee E\}$, since that can be used only for an $\vee$ expression with a finite number of terms. (Recall that a proof must have a finite length.)

$$\frac{\exists x.F(x) \qquad F(x) \vdash A \quad \{x \text{ arbitrary}\}}{A}\{\exists E\}$$

The following theorem gives an example of $\{\exists E\}$. It says that if $P(x)$ always implies $Q(x)$, and also that $P(x)$ holds for some x, then $Q(x)$ also holds for some x.

Theorem 31. $\exists x \,.\, P(x),\ \forall x \,.\, P(x) \to Q(x) \ \vdash\ \exists x \,.\, Q(x)$

Proof.

$$\frac{\exists x \,.\, P(x) \qquad \dfrac{\boxed{P(c)} \qquad \dfrac{\forall x \,.\, P(x) \to Q(x)}{P(c) \to Q(c)}\{\forall E\}}{\dfrac{Q(c)}{}\{\to E\}}}{\dfrac{Q(c)}{\exists x \,.\, Q(x)}\{\exists I\}}\{\exists E\}$$

□

The following theorem says that a $\forall$ directly inside an $\exists$ can be brought outside the $\exists$.

Theorem 32. $\exists x.\ \forall y.\ F(x,y) \vdash \forall y.\ \exists x.\ F(x,y)$

Proof.

$$
\cfrac{
\exists x.\ \forall y.\ F(x,y) \qquad
\cfrac{
\cfrac{
\cfrac{
\cfrac{\boxed{\forall y.\ F(p,y)}}{F(p,q)}\{\forall E\}
}{\exists x.\ F(x,q)}\{\exists I\}
}{\forall y.\ \exists x.\ F(x,y)}\{\forall I\}
}{}
}{\forall y.\ \exists x.\ F(x,y)}\{\exists E\}
$$

$\qquad\qquad\qquad\qquad\qquad\qquad\qquad\qquad\qquad\qquad\qquad$ $\square$

Exercise 11. The converse of Theorem 32 is the following:

$$\forall y.\ \exists x.\ F(x,y) \vdash \exists x.\ \forall y.\ F(x,y) \qquad \textbf{\textit{Wrong!}}$$

Give a counterexample that demonstrates that this statement is *not* valid.

Exercise 12. Prove $\forall x.(F(x) \wedge G(x)) \vdash (\forall x.F(x)) \wedge (\forall x.G(x))$.

3.4   Algebraic Laws of Predicate Logic

The previous section presented predicate logic as a natural deduction inference system. An alternative style of reasoning is based on a set of algebraic laws about propositions with predicates, listed in Table 3.1.

This is not the minimal possible set of laws; some of them correspond to inference rules, and others are provable as theorems. The focus in this section, however, is on practical calculations using the laws, rather than on theoretical foundations.

The following two laws express, in algebraic form, the $\{\forall E\}$ and $\{\exists I\}$ inference rules. Since they correspond to inference rules, these laws are logical implications, not equations.

$$\forall x.\ f(x) \quad \rightarrow \quad f(c) \tag{3.3}$$
$$f(c) \quad \rightarrow \quad \exists x.\ f(x) \tag{3.4}$$

In both of these laws, x is bound by the quantifier, and it may be any element of the universe. The element c is any fixed element of the universe. Thus the first law says that if the predicate f holds for all elements of the universe, it must hold for a particular one c, and the second law says that if f holds for an *arbitrarily chosen* element c then it must hold for all elements of the universe.

The following theorem combines these two laws, and is often useful in proving other theorems. Its proof uses the line-by-line style which is standard when reasoning about predicate logic with algebraic laws.

Table 3.1: Algebraic Laws of Predicate Logic

$$
\begin{array}{rcll}
\forall x.\, f(x) & \to & f(c) & (3.3) \\
f(c) & \to & \exists x.\, f(x) & (3.4) \\[6pt]
\forall x.\, \neg f(x) & = & \neg \exists x.\, f(x) & (3.5) \\
\exists x.\, \neg f(x) & = & \neg \forall x.\, f(x) & (3.6) \\[6pt]
\forall x.\, f(x) \wedge q & = & \forall x.\, (f(x) \wedge q) & (3.7) \\
\forall x.\, f(x) \vee q & = & \forall x.\, (f(x) \vee q) & (3.8) \\
\exists x.\, f(x) \wedge q & = & \exists x.\, (f(x) \wedge q) & (3.9) \\
\exists x.\, f(x) \vee q & = & \exists x.\, (f(x) \vee q) & (3.10) \\[6pt]
\forall x.\, f(x) \wedge \forall x.\, g(x) & = & \forall x.\, (f(x) \wedge g(x)) & (3.11) \\
\forall x.\, f(x) \vee \forall x.\, g(x) & \to & \forall x.\, (f(x) \vee g(x)) & (3.12) \\
\exists x.\, (f(x) \wedge g(x)) & \to & \exists x.\, f(x) \wedge \exists x.\, g(x) & (3.13) \\
\exists x.\, f(x) \vee \exists x.\, g(x) & = & \exists x.\, (f(x) \vee g(x)) & (3.14 \\
\end{array}
$$

Theorem 33. $\forall x.\, f(x) \to \exists x.\, f(x)$

Proof.

$$
\begin{aligned}
& \forall x.\, f(x) \\
\to\ & f(c) && \{3.3\} \\
\to\ & \exists x.\, f(x) && \{3.4\}
\end{aligned}
$$

☐

The next two laws state how the quantifiers combine with logical negation. The first one says that if $f(x)$ is always false, then it is never true; the second says that if $f(x)$ is ever untrue, then it is not always true.

$$
\begin{array}{rcll}
\forall x.\, \neg f(x) & = & \neg \exists x.\, f(x) & (3.5) \\
\exists x.\, \neg f(x) & = & \neg \forall x.\, f(x) & (3.6)
\end{array}
$$

The following four laws show how a predicate $f(x)$ combines with a proposition q that does not contain x. These are useful for bringing constant terms into or out of quantified expressions.

$$
\begin{array}{rcll}
\forall x.\, f(x) \wedge q & = & \forall x.\, (f(x) \wedge q) & (3.7) \\
\forall x.\, f(x) \vee q & = & \forall x.\, (f(x) \vee q) & (3.8) \\
\exists x.\, f(x) \wedge q & = & \exists x.\, (f(x) \wedge q) & (3.9) \\
\exists x.\, f(x) \vee q & = & \exists x.\, (f(x) \vee q) & (3.10)
\end{array}
$$

The final group of laws concerns the combination of quantifiers with $\land$ and $\lor$. It is important to note that two of them are equations (or double implications), while the other two are implications and can be used in only one direction.

$$\forall\, x.\, f(x) \land \forall\, x.\, g(x) \quad = \quad \forall\, x.\, (f(x) \land g(x)) \qquad (3.11)$$
$$\forall\, x.\, f(x) \lor \forall\, x.\, g(x) \quad \rightarrow \quad \forall\, x.\, (f(x) \lor g(x)) \qquad (3.12)$$
$$\exists\, x.\, (f(x) \land g(x)) \quad \rightarrow \quad \exists\, x.\, f(x) \land \exists\, x.\, g(x) \qquad (3.13)$$
$$\exists\, x.\, f(x) \lor \exists\, x.\, g(x) \quad = \quad \exists\, x.\, (f(x) \lor g(x)) \qquad (3.14)$$

Example 9. The following equation can be proved algebraically:

$$\forall x.\, (f(x) \land \neg g(x)) \;=\; \forall x. f(x) \;\land\; \neg \exists x.\, g(x)$$

This is established through a sequence of steps. Each step should be justified by one of the algebraic laws, or by another equation that has already been proved. When the purpose is actually to prove a theorem, the justifications should be written explicitly. Often this kind of reasoning is used informally, like a straightforward algebraic calculation, and the formal justifications are sometimes omitted.

$$
\begin{aligned}
& \forall x.\, (f(x) \land \neg g(x)) \\
={} & \forall x.\, f(x) \;\land\; \forall x.\, \neg g(x) \quad \{3.11\} \\
={} & \forall x. f(x) \;\land\; \neg \exists x.\, g(x) \quad \{3.5\}
\end{aligned}
$$

Example 10. The following equation says that if $f(x)$ sometimes implies $g(x)$, and $f(x)$ is always true, then $g(x)$ is sometimes true.

$$\exists x.\, (f(x) \rightarrow g(x)) \;\land\; (\forall x.\, f(x)) \;=\; \exists x.\, g(x)$$

The first step of the proof replaces the local variable x by y in the $\forall$ expression. This is not actually necessary, but it may help to avoid confusion; whenever the same variable is playing different roles in different expressions, and there seems to be a danger of getting them mixed up, it is safest just to change the local variable. In the next step, the $\forall$ expression is brought inside the $\exists$; in the following step it is now possible to pick a particular value for y: namely the x bound by $\exists$.

$$
\begin{aligned}
& \exists x.\, (f(x) \rightarrow g(x)) \;\land\; (\forall x.\, f(x)) \\
={} & \big(\exists x.\, (f(x) \rightarrow g(x))\big) \;\land\; (\forall y.\, f(y)) && \text{change of variable} \\
={} & \exists x.\, \big((f(x) \rightarrow g(x)) \;\land\; (\forall y.\, f(y))\big) && \{3.9\} \\
={} & \exists x.\, \big((f(x) \rightarrow g(x)) \;\land\; f(x)\big) && \{3.3\} \\
={} & \exists x.\, g(x) && \{Modus\ Ponens\}
\end{aligned}
$$

3.5 Suggestions for Further Reading

The books on logic recommended at the end of Chapter 2 also cover predicate logic. Those citations are not repeated here; instead, two excellent books on style and elegance in mathematical proofs are suggested.

Mathematical Writing, by Knuth, Larrabee and Roberts [19], is filled with good advice about how to write mathematics in a style which is clear, rigorous and lively. This short book is based on a course on mathematical writing given by the authors at Stanford.

A *rigorous* proof is careful, and it covers scrupulously all the relevant aspects of the problem. Routine, straightforward issues may be treated lightly in a rigorous proof, but they really must be sound. In contrast, a *formal* proof takes no short cuts at all; it does *everything* using the rules of some formal system, such as the logical inference rules. Formal proofs are good for machine checking, and generally fit well with computing applications. Proofs that are informal but rigorous should be written in a clear, elegant style that is convincing to a knowledgeable reader.

Proofs from THE BOOK, by Aigner and Ziegler [2], is an outstanding collection of elegant and rigorous proofs written in normal (but particularly good) mathematical style. It is worth looking at for its beauty, although some of its contents are rather advanced. The book was inspired by an idea of Paul Erdős, one of the leading mathematicians of the twentieth century. Erdős imagined The Book, which contains the most elegant proofs of the most interesting theorems. The Book doesn't actually exist; it is an ideal to which real people can only aspire, but it is nevertheless inspiring to mathematicians to find the best approximation to it which they can. *Proofs from THE BOOK* is such an effort, and it is likely to become a mathematical classic.

3.6 Review Exercises

Exercise 13. Suppose the universe contains 10 elements. How many times will F occur when $\forall x.\ \exists y.\forall z.\ F(x, y, z)$ is expanded into quantifier-free form? How large in general are expanded expressions?

Exercise 14. Prove $(\exists x.\ f(x)) \lor (\exists x.\ g(x)) \vdash \exists x.\ (f(x) \lor g(x))$.

Exercise 15. Prove $(\forall x.\ f(x)) \lor (\forall x.\ g(x)) \vdash \forall x.\ (f(x) \lor g(x))$.

Exercise 16. Prove the converse of Theorem 29.

Exercise 17. Find counterexamples which show that Laws 3.12 and 3.13, which are implications, would not be valid as equations.

Exercise 18. Prove the following implication:

$$\big(\forall x.\ f(x) \to h(x)\ \land\ \forall x.\ g(x) \to h(x)\big)$$
$$\to\quad \forall x.\ (f(x) \lor g(x) \to h(x))$$

Chapter 4

Set Theory

Set theory is one of the most fundamental branches of mathematics. Many profound advances in mathematics over the last century have taken place in set theory, and there is a deep connection between set theory and logic. More importantly for computer science, it has turned out that the notation and terminology of elementary set theory is extremely useful for describing algorithms, and nearly every branch of computing uses sets from time to time.

This chapter introduces the concepts from set theory that you will need for computer science. Section 4.1 begins by describing what sets are and giving several notations for describing them. There are many useful operations that can be performed on sets, and these are presented in Section 4.2. In Section 4.3 we consider a particular kind of set that is particularly well suited for computing applications, the finite sets with equality. Next, in Section 4.4 we study a variety of mathematical laws that describe properties of sets that are useful in computing. The chapter concludes with a summary of the notations and the main theorems of set theory.

4.1 Notations for Describing Sets

We will not define formally what a set is. The reason for this is that set theory was intended originally to serve as the foundation for all of mathematics: everything else in mathematics is to be defined—at least in principle—in terms of sets. Since sets are the lowest level concept, there is nothing more primitive that could be used to define them formally.

Informally, a set is just a collection of objects called *members* or *elements*. You can think of a set as a group of members, or a collection, or a class, etc. However, these words are just synonyms—they don't constitute a precise definition of a set. One way to describe a set is to write down all its members inside braces { }. Here are some examples:

111

$$
\begin{aligned}
A &= \{\text{dog}, \text{cat}, \text{horse}\} \\
B &= \{\text{canary}, \text{eagle}\} \\
C &= \{0, 1, 2, 3, 4\} \\
D &= \{0, 1, \dots, 100\} \\
E &= \{\,\} \\
N &= \{0, 1, 2, 3, \dots\} \\
Z &= \{3, \{\text{dog}, 7\}, \text{horse}\}
\end{aligned}
$$

Any particular thing can appear only once in a set; this means that it makes sense to ask whether x is a member of a set S—the answer must be yes or no—but it doesn't make sense to ask how many times x occurs in S. It is bad notation to write a set with some element appearing several times, because the extra occurrences of the element are meaningless, and they might be confusing. You can always remove the redundant copies of an element without changing the set.

A set can have any number of elements. For example, A has three elements, E has zero elements, and N has an infinite number of elements.

It is common to use lower case letters $(a, b, \dots)$ to refer to members of a set, and to use upper case letters $(A, B, \dots)$ as names for sets themselves. This is just a convention, not an ironclad rule, and of course it breaks down when one set is a member of another set.

An important special case is the empty set $\{\,\}$. Often the special symbol $\emptyset$ is used to denote the empty set.

Suppose we are given some value x and a set S, and we want to know whether x is a member of S. There is a notation for this question: the expression $x \in S$ is true if x is a member of S and otherwise false. The expression $x \in S$ is pronounced as 'x is a member of S', or 'x is an element of S', or simply as 'x is in S'. For example:

$$
\begin{aligned}
\text{dog} \in A &= \text{True} \\
\text{bat} \in A &= \text{False}
\end{aligned}
$$

Another useful notation is $x \notin S$, which is True if and only if $x \in S$ is False:

$$
\begin{aligned}
\text{dog} \notin A &= \text{False} \\
\text{bat} \notin A &= \text{True}
\end{aligned}
$$

When a set has a few elements, you can just write them out inside braces. This is how the sets A, B and C were defined above. However, when a set has many elements this becomes tedious, and if the set has an infinite number of elements it is impossible. The set D has 101 elements, but it is more readable to use the ... notation and omit most of them. Set N, the natural numbers, has an infinite number of elements and we cannot write them all inside braces.

There is an interesting point about the ... notation. One of the reasons we use mathematics in computing is to be precise and formal, to be absolutely sure there is no ambiguity in what we are defining. The ... notation is informal, and it relies on the intuition of the reader to understand and fill in the dots. It is easy to construct cases where different people might interpret a set defined with ... differently. For example, does $\{2, 3, \dots, 7\}$ mean the set of numbers from 2 through 7, $\{2, 3, 4, 5, 6, 7\}$, or does it mean the set of prime numbers $\{2, 3, 5, 7\}$? If you are aware of such a problem when describing a set, you can overcome it, but how can you ever be sure that your set is really well-defined—that there is one way, and only one way, to interpret it? The problem is not so serious for finite sets, because you could—in principle—write out all the elements, and for large sets you could provide an algorithm that produces all of them. The problem is more fundamental for infinite sets.

Another standard way to define sets is the *set comprehension*. In its simplest form, a set comprehension is written as

$$\{x \mid p\ x\},$$

where $p\ x$ is simply an expression containing x which is either true or false; such an expression is called a *predicate*. The expression is pronounced 'the set of x such that $p\ x$', and it means that the set consists of exactly those objects x of which $p\ x$ is true. For example, we could define the set of even numbers as

$$\{x \mid x \in N \wedge \text{even } x\}.$$

The predicate here is

$$p\ x\ =\ x \in N\ \wedge\ \text{even } x,$$

and it is true if and only if x is a natural number that is even. In English, we would call this 'the set of all x such that x is a natural number and x is even', or simply 'the set of even natural numbers'.

A more general form of the set comprehension is

$$\{f\ x \mid p\ x\}.$$

In this case, the set consists not of the values x that satisfy the predicate, but of the results of applying the function f to those values. This form of the set comprehension is sometimes easier to use than the simpler form. For example

$$\{\sqrt{x} \mid x \in \{1, 2, 3, 4\}\}$$

defines the set $\{1.0, 1.41, 1.73, 2.0\}$.

It is important to state the set from which a variable derives its value. If this set is not stated explicitly, then we assume that it is U, the universe of discourse.

4.2   Basic Operations on Sets

There is a large number of operations that can be performed on sets, in order
to compare them, define new sets and so on. This section defines the basic set
operations. In the next section we will see how to implement these operations
on a computer.

4.2.1   Subsets and Set Equality

There are several important relationships between two sets that are determined
by the elements they share. The first of these is the *subset* relation. The
expression $A \subseteq B$, pronounced 'A is a subset of B', is true if each element of A
also appears in B. This idea is expressed formally by the following definition.

Definition 11 (Subset). Let A and B be sets. Then $A \subseteq B$ if and only if

$$\forall x. x \in A \rightarrow x \in B.$$

Two sets are equal if they contain exactly the same elements. We can define
this formally using the subset relation, since if A and B contain exactly the
same elements, then everything in A is also in B and vice versa. This leads to
the definition of set equality.

Definition 12 (Set equality). Let A and B be sets. Then $A = B$ if and
only if $A \subseteq B$ and $B \subseteq A$.

If A is a subset of B but $A \neq B$, then all the elements of A are in B but
there must be some element of B that is not in A. In this case we say that A is
a *proper subset* of B, which is written as $A \subset B$. The notations are designed to
help you remember them. Think of $A \subset B$ as saying that the set A is contained
within B, and it is smaller than B, while $A \subseteq B$ means that possibly $A = B$;
the symbols $\subset$ and $\subseteq$ are reminiscent of $<$ and $\leq$.

Definition 13 (Proper subset). Let A and B be sets. Then $A \subset B$ if and
only if $A \subseteq B$ and $A \neq B$.

4.2.2   Union, Intersection and Difference

There are several operators that take two sets and return a set as a result; the
most important of these are union, intersection and difference.

- The *union* of two sets A and B, written $A \cup B$, is the set that contains
  all the elements that are in either A or B (or both). Every element of
  $A \cup B$ must be in A or B (or both).

- The *intersection* of A and B, written $A \cap B$, is the set consisting of all
  the elements that are in *both* A and B.

- The *difference* of A and B, written A − B, is the set of all the elements that are in A but not in B.

Definition 14. Let A and B be sets. Then

$$A \cup B \quad = \quad \{x \mid x \in A \vee x \in B\}, \tag{4.1}$$
$$A \cap B \quad = \quad \{x \mid x \in A \wedge x \in B\}, \tag{4.2}$$
$$A - B \quad = \quad \{x \mid x \in A \wedge x \notin B\}. \tag{4.3}$$

Example 11. Let $A = \{1, 2, 3\}$ and $B = \{3, 4, 5\}$. Then

$$A \cup B \quad = \quad \{1, 2, 3, 4, 5\},$$
$$A \cap B \quad = \quad \{3\},$$
$$A - B \quad = \quad \{1, 2\}.$$

Example 12. Let

$$I \quad = \quad \{\ldots, -2, -1, 0, 1, 2, \ldots\},$$
$$N \quad = \quad \{0, 1, 2, \ldots\},$$
$$H \quad = \quad \{-2^{15}, \ldots, -2, -1, 0, 1, 2, \ldots, 2^{15} - 1\},$$
$$W \quad = \quad \{-2^{31}, \ldots, -2, -1, 0, 1, 2, \ldots, 2^{31} - 1\}.$$

Thus I is the set of integers, N is the set of natural numbers, H is the set of integers that are representable on a computer with a 16 bit word using 2's complement number representation, and W is the set of integers that are representable in a 32 bit word. We can use these definitions to create new sets. For example, $I - W$ is the set of integers that are not representable in a word.

We can calculate the union of several sets using the $\cup$ operator. For example, the union of three sets A, B and C can be written as

$$A \cup B \cup C,$$

and it contains all the elements that appear in one or more of the sets A, B and C. This expression is unambiguous because the $\cup$ operator is associative (see Section 4.4), which means that it makes no difference whether you interpret $A \cup B \cup C$ as $(A \cup B) \cup C$ or as $A \cup (B \cup C)$. In the same way, we can calculate the intersection of four sets with the expression

$$A \cap B \cap C \cap D.$$

Sometimes it is necessary to compute the union (or intersection) of several sets in a more general way, using operators that give the union (or intersection) of an arbitrary number of sets, rather than just two of them. These operations are often called *big union* and *big intersection*, because their operators, $\bigcup$ and $\bigcap$, are larger versions of the ordinary $\cup$ and $\cap$ operators.

Definition 15. Let C be a non-empty collection (set) of subsets of the universe U. Let I be a non-empty set, and for each $i \in I$ let $A_i \subseteq C$. Then

$$\bigcup_{i \in I} A_i \;=\; \{x \mid \exists i \in I \,.\, x \in A_i\},$$

$$\bigcap_{i \in I} A_i \;=\; \{x \mid \forall i \in I \,.\, x \in A_i\}.$$

Another way to say this is that if C is a set containing some sets, then the set of all elements of the sets in C is $\bigcup_{A \in C} A$ and the set of elements which these sets in C have in common is $\bigcap_{A \in C} A$.

$$\bigcup_{A \in C} A \;=\; \{x \mid \exists A \in C \,.\, x \in A\}$$

$$\bigcap_{A \in C} A \;=\; \{x \mid \forall A \in C \,.\, x \in A\}$$

Two sets are *disjoint* if they have no elements in common.

Definition 16. For any two sets A and B, if $A \cap B = \emptyset$ then A and B are *disjoint* sets.

Exercise 1. Given the sets $A = \{1, 2, 3, 4, 5\}$ and $B = \{2, 4, 6\}$, calculate the following sets:

(a) $A \cup B \cap A$

(b) $(A \cap B) \cup B$

(c) $A - B$

(d) $(B - A) \cap B$

(e) $A \cup (B - A)$

4.2.3   Complement and Power

In many applications there is a *universe* of all the objects that might possibly appear in any of our sets. For example, we might be working with various sets of numbers, but none of the sets will contain anything that is *not* a number. In this kind of situation it is often convenient to define the universe explicitly as a set U, which can then be used in set expressions.

The universe is needed to define the*complement* of a set. The intuitive idea is that the complement of a set A is the set of everything that is *not* in A. However, what does 'everything' mean? There are both practical and theoretical problems if we don't define 'everything'. A practical problem is that we could get nonsensical results; for example, if we are talking about sets of people, then the complement of the set of tall people should be the set that includes short people—but not numbers, cars and toasters, all of which appear in the set of 'everything'. The solution to these problems is to define the universe U to consist of the elements we are interested in, and then to define the complement of A to consist of the elements of U that are not in A.

Definition 17. Let U be the universe of discourse and A be a set. The *complement* of A, written A', is the set $U - A$.

If you see a set complement defined in a book or paper, look back several pages and you should find the definition of U. When you're using sets, be sure to define the universe explicitly if you are going to use complements.

Example 13. Given the universe of alphanumeric characters, the complement of the set of digits is the set of letters.

Example 14. If the universe is $\{1, 2, 3, 4, 5\}$, then $\{1, 2\}' = \{3, 4, 5\}$.

A set that contains lots of elements will have an even larger number of subsets. These subsets are themselves objects, and it is often useful to define a new set containing all of them. The set of all subsets of A is called the *powerset* of A. (In contrast, the set of all *elements* of A is just A itself.)

Definition 18. Let A be a set. The *powerset* of A, written $P(A)$, is the set of all subsets of A:

$$P(A) = \{S \mid S \subseteq A\}$$

Example 15. (Powersets)

- $P(\{\,\}) = \{\emptyset\} = \{\{\,\}\}$
- $P(\{a\}) = \{\emptyset, \{a\}\}$
- $P(\{a, b\}) = \{\emptyset, \{a\}, \{b\}, \{a, b\}\}$
- $P(\{a, b, c\}) = \{\emptyset, \{a\}, \{b\}, \{c\}, \{a, b\}, \{a, c\}, \{b, c\}, \{a, b, c\}\}$

Notice that if A contains n elements, then its powerset $P(A)$ contains 2^n elements. If A is the empty set $\emptyset$, then there is one element of $P(A)$—namely $\emptyset$—and the number of elements of $P(A)$ is indeed $1 = 2^0$. The example above shows that sets with 0, 1, 2 and 3 elements have powersets respectively containing 1, 2, 4 and 8 elements.

4.3   Finite Sets with Equality

A class of sets that is particularly important for computing are the finite sets with equality—that is, sets with a finite number of elements, and where we have an equality function that can be used to determine whether any two elements of the universe are the same.

It is possible to perform some computations with infinite sets, but there is ample opportunity for a program to go into an infinite loop. Usually we want to ensure that our programs will terminate, and this is more straightforward with finite sets.

We will use lists to represent sets; thus a set whose elements are of type a will be represented by a list of type $[a]$. A finite set will be represented by a finite list. There are several differences between lists and sets which will require care: (1) lists may have duplicate elements, (2) there is a fixed order of the elements of a list, and (3) all the elements of a list must have the same type.

In order to do any practical computing with sets, we need to be able to determine whether a given value is an element of a set. This requires, in turn, the ability to determine whether two values x and y are the same. Therefore we need an operator $==$ so that $x == y$ will give either True or False, for any values of x and y. This is a strong requirement, because there are some values where equality is not computable. For all the elementary data types, such as integers, booleans, characters, and so on, there is no problem with determining equality. However, functions are also values that can be included in sets, and it is in general impossible to compare two functions to determine whether they are equal.

There is a special notation in Haskell to express the fact that it must be possible to compare two set elements for equality. If we say simply that a set is represented by a list of type $[a]$, it might turn out that a is one of those types whose values cannot be compared. What is needed is a way to restrict the element type a. This is done by saying that a list has type

$$Eq\ a \Rightarrow [a].$$

This denotes the type 'For every type a which can be compared for Equality, the type $[a]$'.

A set A will be represented by a list containing all the elements of A. We need to be careful in computing with such lists because of two factors: the possibility of duplicate elements and the ordering of the elements.

A set contains just one instance of each of its elements; thus the sets $\{1, 2, 3\}$ and $\{1, 2, 1, 3\}$ are identical (and the second way of writing it is bad notation because 1 appears only once in the set; it should appear only once in the notation). However, the lists $[1, 2, 3]$ and $[1, 2, 1, 3]$ are different. The first one has three elements and the second has four elements. We will assume that a list represents the set of distinct values that appear in the list: for example, the list $[1, 2, 1, 3]$ represents the set $\{1, 2, 3\}$. However, the *normal form* for a list representing a set will contain each element only once. The operations on set-representing-lists will all produce their result in normal form. This makes some of the operations easier to implement, and also makes them more efficient.

There is no concept of ordering of the elements of a set. Thus $\{1, 2, 3\}$ and $\{3, 2, 1\}$ are the same set. If you print a set, there is no particular order in which the elements will appear. Lists, however, have a specific ordering of their elements; thus $[1, 2, 3]$ and $[3, 2, 1]$ are different lists and their elements will be printed in different orders. This does not cause any difficulties for implementing the set operations. In a computer program, however, it may be a good idea to define an ordering on the elements of a set in order to make the output more readable. This can be achieved by sorting the elements of the list that represents the set. However, sorting a list requires more than the ability to compare two elements for equality: we must also be able to compare them for ordering ($<, =, >$). If a program uses sorted lists to represent sets, then a stronger type constraint is needed:

$$Ord\ a \Rightarrow [a].$$

This says that there must be an ordering on the element type a, which can be used to determine the relations $<, \leq, =, \neq, >, \geq$.

The methods for defining lists can also be used to define sets. Three methods are especially common: enumerated sets, sequences, and comprehensions. We will look at each of these methods in turn, and the following section defines functions corresponding to the set operators we have already covered.

An enumerated set (or list) is defined just by giving its elements, like $\{1, 2, 3\}$ and $[1, 2, 3]$.

A *sequence* is used when it would be too tedious to write an enumerated set or list. For example, the set of natural numbers up to 5 is $\{0, 1, 2, 3, 4, 5\}$, but it would be painful to write the set of natural numbers up to 1,000. Instead, the '…' notation is used in mathematics: $\{0, 1, 2, \ldots, 1000\}$. We can use a similar notation to define the corresponding list: $[0, 1..1000]$.

Sets are often described in mathematics using *set comprehensions*, expressions like

$$\{x^2 \mid x \in \{0, 1, \ldots, n\}\},$$

which is the set of numbers of the form x^2 where x is a natural number between 0 and n. Haskell provides an almost identical notation, the *list comprehension*. A list comprehension that expresses the set just given is:

```
[x^2 | x <- [0..n]]
```

Here are some other examples:

```
[ x | x <- [0..n], x 'mod' 2 == 0]
[ x + 3 | x <- [0..n], x < 10]
```

4.3.1 Computing with Sets

This section describes a collection of functions and operators that you can use on finite sets with equality. Each of these functions will accept a representation with duplicated elements, but they always return results in normal form; that is, they remove any duplicated elements from their result.

We begin by defining the type of set representations.

```
type Set a = [a]
```

The function `normalForm` determines whether there are any duplicate elements within its argument, while `normalizeSet` takes a list and removes any duplicate elements.

```
normalForm :: (Eq a, Show a) => [a] -> Bool
normalizeSet :: Eq a => Set a -> Set a
```

Example 16. `normalForm [1,2,3]` is True, but `normalForm [1,2,1,3]` is False because there is a repeated element, and `normalizeSet [1,2,1,3]` removes the repeated element, returning [1,2,3].

We define symbolic operators for the set operations that take two arguments:

$$
\begin{aligned}
\text{A+++B} &= A \cup B & (union) \\
\text{A***B} &= A \cap B & (intersection) \\
\text{A}\sim\sim\sim\text{B} &= A - B & (difference)
\end{aligned}
$$

The types of these operators are:

```
(+++) :: (Eq a, Show a) => Set a -> Set a -> Set a
(***) :: (Eq a, Show a) => Set a -> Set a -> Set a
(~~~) :: (Eq a, Show a) => Set a -> Set a -> Set a
```

The function `subset` takes two sets A and B, and returns True if $A \subseteq B$, while `properSubset A B` returns True if A is a proper subset of B; that is, it returns the value of $A \subset B$. Their type signatures are:

```
subset, properSubset ::
    (Eq a, Show a) => Set a -> Set a -> Bool
```

A special function `setEq` is needed to determine whether two sets are the same. The built-in Haskell operator `==` should not be used for this purpose, since two lists may be different even though they represent the same set.

```
setEq :: (Eq a, Show a) => Set a -> Set a -> Bool
```

The complement of a set is defined with respect to the universe of discourse. It is convenient to define a global variable `universe` to be the universe of discourse; then we can define the set `complement` function as follows:

```
complement s = universe ~~~ s
```

Complement with respect to a specific universe is defined as `!!!` in Haskell rather than `complement`.

```
(!!!) :: (Eq a, Show a) => Set a -> Set a -> Set a
```

Exercise 2. Work out the values of the following set expressions, and then check your answer using the Haskell expression that follows.

(a) [1,2,3] +++ [3]

(b) [4,2] +++ [2,4]

(c) [1,2,3] *** [3]

(d) [] *** [1,3,5]

(e) [1,2,3] ~~~ [3]

(f) [2,3] ~~~ [1,2,3]

(g) [1,2,3] *** [1,2]

(h) [1,2,3] +++ [4,5,6]

(i) ([4,3] ~~~ [5,4]) *** [1,2]

(j) ([3,2,4] +++ [4,2]) ~~~ [2,3]

(k) subset [3,4] [4,5,6]

(l) subset [1,3] [4,1,3,6]

(m) subset [] [1,2,3]

(n) setEq [1,2] [2,1]

(o) setEq [3,4,6] [2,3,5]

(p) [1,2,3] !!! [1]

(q) [] !!! [1,2]

Exercise 3. The function

```
powerset :: (Eq a, Show a) => Set a -> Set (Set a)
```

takes a set and returns its power set. Work out the values of the following expressions:

```
powerset [3,2,4]
powerset [2]
```

Exercise 4. The function

```
crossproduct :: (Eq a, Show a, Eq b, Show b) =>
                Set a -> Set b -> Set (a,b)
```

takes two sets and returns their cross product. Evaluate these expressions:

```
crossproduct [1,2,3] ['a','b']
crossproduct [1] ['a','b']
```

Exercise 5. In the following exercise, let u be [1,2,3,4,5,6,7,8,9,10], a
be [2,3,4], b be [5,6,7] and c be [1,2]. Give the elements of each set:

```
a +++ b
u!!!a *** (b +++ c)
c ~~~ b
(a +++ b) +++ c
u!!!a
u!!!(b *** c)
```

Exercise 6. What are the elements of the set $\{x+y \mid x \in \{1,2,3\} \wedge y \in \{4,5\}\}$?

Exercise 7. Write and evaluate a list comprehension that expresses the set
$\{x \mid x \in \{1,2,3,4,5\} \wedge x < 0\}$

Exercise 8. Write and evaluate a list comprehension that expresses the set
$\{x + y \mid x \in \{1,2,3\} \wedge y \in \{4,5\}\}$

Exercise 9. Write and evaluate a list comprehension that expresses the set
$\{x \mid x \in \{1,2,3,4,5,6,7,8,9,10\} \wedge even\ x\}$

Exercise 10. What is the value of each of the following expressions?

```
subset [1,3,4] [4,3]
subset [] [2,3,4]

setEq [2,3] [4,5,6]
setEq [1,2] [1,2,3]
```

4.4   Set Laws

The operations on sets that we have been covering are often used for describing
properties of algorithms, so we often need to be able to understand and calculate
with expressions that contain several such operations. Fortunately, the set
operations satisfy a number of simple laws that greatly simplify their use, just
as the basic properties of ordinary addition, multiplication, etc. are helpful in
ordinary algebra. In this section we state and prove some of the most useful laws
about set operations. The proofs have a standard form in which an assertion
appears, followed by its justification, which may depend on previous lines in
the proof and cite their line numbers. As justifications take many forms in
practice, these will be more terse than those in previous proofs.

The following law says that the $\subseteq$ operation is transitive:

Theorem 34. Let A, B, and C be sets. If $A \subseteq B$ and $B \subseteq C$, then $A \subseteq C$.

Proof. Let x be any element of the universe of discourse.

1.	$A \subseteq B$	Premise
2.	$x \in A \to x \in B$	Def. $\subseteq$
3.	$B \subseteq C$	Premise
4.	$x \in B \to x \in C$	Def. $\subseteq$
5.	$x \in A \to x \in C$	Hypothetical syllogism, (2),(4)
6.	$\forall x. (x \in A \to x \in C)$	$\forall$ introduction
7.	$A \subseteq C$	Def. $\subseteq$

□

Exercise 11. Let A, B, and C be sets. Prove that if $A \subset B$ and $B \subset C$, then $A \subset C$.

Exercise 12. Consider the following two claims. For each one, if it is true give a proof, but if it is false give a counterexample.

(a) If $A \subseteq B$ and $B \subseteq C$, then $A \subset C$.

(b) If $A \subset B$ and $B \subset C$, then $A \subseteq C$.

4.4.1 Associative and Commutative Set Operations

The set union and intersection operators are commutative and associative.

Theorem 35. For all sets A, B and C,

1. $A \cup B = B \cup A$

2. $A \cap B = B \cap A$

3. $A \cup (B \cup C) = (A \cup B) \cup C$

4. $A \cap (B \cap C) = (A \cap B) \cap C$

5. $A - B = A \cap B'$

Proof. We prove the second equation. Let x be any element of U. Then:

1.	$x \in A \cap B$	Premise
2.	$x \in A \wedge x \in B$	Def. $\cap$
3.	$x \in B \wedge x \in A$	Comm. $\wedge$
4.	$x \in B \cap A$	Def. $\cap$
5.	$\forall x \in U.$	
	$x \in A \cap B \leftrightarrow x \in B \cap A$	Pred. logic abstraction
6.	$A \cap B = B \cap A.$	Def. set equality

□

The proofs of the other equations are similar.

4.4.2 Distributive Laws

The following theorem states that the union and intersection operators distribute over each other.

Theorem 36. $A \cap (B \cup C) = (A \cap B) \cup (A \cap C)$

Proof. Let x be an arbitrary element of the universe U. Then:

1. $x \in A \cap (B \cup C)$ Premise
2. $x \in A \wedge (x \in B \cup C)$ Def. $\cap$
3. $x \in A \wedge (x \in B \vee x \in C)$ Def. $\cup$
4. $(x \in A \wedge x \in B) \vee (x \in A \wedge x \in C)$ Distr. $\wedge$ over $\vee$
5. $(x \in A \cap B) \vee (x \in A \cap C)$ Def. $\cap$
6. $x \in (A \cap B) \cup (A \cap C)$ Def. $\cup$
7. $\forall x \in U.$
    $x \in A \cap (B \cup C) \leftrightarrow x \in (A \cap B) \cup (A \cap C)$ Pred. logic abstraction
8. $A \cap (B \cup C) = (A \cap B) \cup (A \cap C).$ Def. set equality

$\square$

Theorem 37. $A \cup (B \cap C) = (A \cup B) \cap (A \cup C)$

Proof. Exercise for the reader. $\square$

4.4.3 DeMorgan's Laws for Sets

Theorem 38. Let A and B be arbitrary sets. Then

$$(A \cup B)' = A' \cap B'$$

and

$$(A \cap B)' = A' \cup B'.$$

Proof. We prove that $(A \cup B)' = A' \cap B'$. Let x be any element of U. Then:

1. $x \in (A \cup B)'$ Premise
2. $x \in U \wedge \neg(x \in A \cup B)$ Def. comp
3. $x \in U \wedge \neg(x \in A \vee x \in B)$ Def. $\cup$
4. $x \in U \wedge (\neg(x \in A) \wedge \neg(x \in B))$ DeMorgan
5. $x \in U \wedge x \in U \wedge (\neg(x \in A) \wedge \neg(x \in B))$ idemp. of $\wedge$
6. $(x \in U \wedge \neg(x \in A)) \wedge (x \in U \wedge \neg(x \in B))$ comm. of $\wedge$
7. $x \in U - A \wedge x \in U - B$ Def. of diff.
8. $x \in (U - A) \cap (U - B)$ Def. $\cup$
9. $(x \in A' \cap B')$ Def. of comp.
10. $\forall x. \, x \in (A \cup B)' \leftrightarrow x \in (A' \cap B')$ $\forall$ introduction
11. $(A \cup B)' = A' \cap B'$

$\square$

Table 4.1: Summary of Set Notation

Elements	$a, b, c, \ldots$
Sets	$A, B, C, \ldots$
Empty set	$\{\}, \phi$
Enumerated set	$\{e_1, e_2, \ldots\}$
Set comprehension	$\{x \mid \cdots\}$
Cardinality	$\mid A \mid$
Member	$x \in A$
Not member	$x \notin A$
Subset	$A \subseteq B$
Not subset	$A \nsubseteq B$
Proper subset	$A \subset B$
Not proper subset	$A \not\subset B$
Union	$A \cup B$
Intersection	$A \cap B$
Set difference	$A - B$
Cross product	$A \times B$

4.5 Suggestions for Further Reading

Mathematics from the Birth of Numbers, by Gullberg [15], is an interesting general survey of mathematics. It covers many of the topics in this book, including an excellent survey of elementary set theory.

Classic Set Theory [13], by Derek Goldrei, is a self-study textbook telling the full story of set theory, including construction of the real numbers, the Axiom of Choice, cardinal and ordinal numbers, and more. This book is challenging, but it conveys the sense of excitement that surrounded set theory as it was developed.

4.6 Review Exercises

Exercise 13. For the following questions, give a proof using set laws, or find a counterexample.

(a) $(A' \cup B)' \cap C' = A \cap (B \cup C)'$

(b) $A - (B \cup C)' = A \cap (B \cup C)$

(c) $(A \cap B) \cup (A \cap B') = A$

(d) $A \cup (B - A) = A \cup B$

(e) $A - B = B' - A'$

Table 4.2: Set Laws

Idempotent
$$A = A \cup A$$
$$A = A \cap A$$
Domination
$$A \cup U = U$$
$$A \cap \emptyset = \emptyset$$
Identity
$$A \cup \emptyset = A$$
$$A \cap U = A$$
Double Complement
$$A = A''$$
DeMorgan's Laws
$$(A \cup B)' = A' \cap B'$$
$$(A \cap B)' = A' \cup B'$$
Commutative Laws
$$A \cup B = B \cup A$$
$$A \cap B = B \cap A$$
Associative Laws
$$(A \cup B) \cup C = A \cup (B \cup C)$$
$$(A \cap B) \cap C = A \cap (B \cap C)$$
Distributive Laws
$$A \cap (B \cup C) = (A \cap B) \cup (A \cap C)$$
$$A \cup (B \cap C) = (A \cup B) \cap (A \cup C)$$
Absorption Laws
$$A \cup (A \cap B) = A$$
$$A \cap (A \cup B) = A$$

(f) $A \cap (B - C) = (A \cap B) - (A \cap C)$

(g) $A - (B \cup C) = (A - B) \cap (A - C)$

(h) $A \cap (A' \cup B) = A \cap B$

(i) $(A - B') \cup (A - C') = A \cap (B \cap C)$

Exercise 14. Show why a function that finds the simplest form of a set expression might not terminate.

Exercise 15. Simplify the set $A \cup (C \cap A)$.

Exercise 16. Simplify the set $A \cup (C \cap A')$.

Exercise 17. The function

```
smaller :: Ord a => a -> [a] -> Bool
```

takes a value and a list of values and returns True if the value is smaller than the first element in the list. Using this function, write a function that takes a set and returns its powerset. Use `foldr`.

Exercise 18. Prove that $(A \cup B)' = ((A \cup A') \cap A') \cap ((B \cup B') \cap B')$.

Exercise 19. Using a list comprehension, write a function that takes two sets and returns True if the first is a subset of the other.

Exercise 20. What is wrong with this definition of `diff`, a function that takes two sets and returns their difference?

```
diff :: Eq a => [a] -> [a] -> [a]
diff set1 set2 = [e | e <- set2, not (elem e set1)]
```

Exercise 21. What is wrong with this definition of `intersection`, a function that takes two sets and returns their intersection?

```
intersection :: [a] -> [a] -> [a]
intersection set1 set2 = [e | e <- set1, e <- set2]
```

Exercise 22. Write a function using a list comprehension that takes two sets and returns their union.

Exercise 23. Is it ever the case that $A \cup (B - C) = B$?

Exercise 24. Give an example in which $(A \cup C) \cap (B \cup C) = \emptyset$.

Chapter 5

Recursion

Recursion is a *self referential* style of definition commonly used in both mathematics and computer science. It is a fundamental programming tool, particularly important for manipulating data structures.

The next three chapters will look at recursion from several points of view. This chapter introduces the idea by showing how to write recursive functions in computer programs. Chapter 6 applies the same idea to mathematics, where recursion is used to define sets inductively. Finally, Chapter 7 introduces induction, the mathematical technique for proving properties about recursive definitions. You will find examples of recursion and induction throughout many branches of computer science. In addition to the examples given through these three chapters, we will study a larger case study in Chapter 10, where recursion and induction are applied to the problem of digital circuit design.

The idea in recursion is to break a problem down into two cases: if the problem is 'easy' a direct solution is specified, but if it is 'hard' we proceed through several steps: first the problem is redefined in terms of an easier problem; then we set aside the current problem temporarily, and go solve the easier problem; and finally we use the solution to the easier problem to calculate the final result. The idea of splitting a hard problem into easier ones is called the *divide and conquer* strategy, and it is frequently useful in algorithm design.

The factorial function provides a good illustration of the process. Often the factorial function is defined using the clear but informal '...' notation:

$$n! \ = \ 1 \times 2 \times \cdots \times n$$

This definition is fine, but the '...' notation relies on the reader's understanding to see what it means. An informal definition like this isn't well suited for proving theorems, and it isn't an executable program in most programming languages[1]. A more precise recursive definition of factorial consists of the

---

[1] Haskell is a very high level programming language, and it actually allows this style: in Haskell you can define `factorial n = product [1..n]`. Most programming languages,

following pair of equations:

$$0\,! \;\; = \;\; 1$$
$$(n+1)\,! \;\; = \;\; (n+1) \times n\,!$$

In Haskell, this would be written as:

```
factorial :: Int -> Int
factorial 0 = 1
factorial (n+1) = (n+1) * factorial n
```

The first equation specifies the easy case, where the argument is 0 and the answer 1 can be supplied directly. The second equation handles the hard case, where the argument is of the form $n + 1$. First the function chooses a slightly easier problem to solve, which is of size n; then it solves for $n!$ by evaluating `factorial n`, and it multiplies the value of $n!$ by $n + 1$ to get the final result.

Recursive definitions consist of a collection of equations that state properties of the function being defined. There are a number of algebraic properties of the factorial function, and one of them is used as the second equation of the recursive definition. In fact, the definition doesn't consist of a set of commands to be obeyed; it consists of a set of true equations describing the salient properties of the function being defined.

Programming languages that allow this style of definition are often called *declarative languages*, because a program consists of a set of declarations of properties. The programmer must find a suitable set of properties to declare, usually via recursive equations, and the programming language implementation then finds a way to solve the equations. The opposite of a declarative language is an *imperative* language, where you give a sequence of commands which, when obeyed, will result in the computation of the result.

5.1   Recursion Over Lists

For recursive functions on lists, the 'easy' case is the empty list [] and the 'hard' case is a non-empty list, which can be written in the form (x:xs). A recursive function over lists has the following general form:

```
f ::   [a] ->  type of result
f [] =  result for empty list
f (x:xs) =  result defined using (f xs) and x
```

A simple example is the `length` function, which counts the elements in a list; for example, `length [1,2,3]` is 3.

---

however, do not allow this, and even Haskell treats the [1..n] notation as a high level abbreviation which is executed internally using recursion.

```
length :: [a] -> Int
length []    = 0
length (x:xs) = 1 + length xs
```

An empty list contains no elements, so `length []` returns 0. When the list contains at least one element, the function calculates the length of the rest of the list by evaluating `length xs`, and then adds one to get the length of the full list. We can work out the evaluation of `length [1,2,3]` using equational reasoning. This process is similar to solving an algebra problem, and the Haskell program performs essentially the same calculation when it executes. Simplifying an expression by equational reasoning is the right way to 'hand-execute' a functional program. In working through this example, recall that `[1,2,3]` is a shorthand notation for `1:(2:(3:[]))`.

```
length [1,2,3]
= 1 + length [2,3]
= 1 + (1 + length [3])
= 1 + (1 + (1 + length []))
= 1 + (1 + (1 + 0))
= 3
```

It is better to think of recursion as a systematic calculation, as above, than to try to imagine low-level subroutine operations inside the computer. Textbooks on imperative programming languages sometimes explain recursion by resorting to the underlying machine language implementation. In a functional language, recursion should be viewed at a high level, as an equational technique, and the low level details should be left to the compiler.

The function `sum` provides a similar example of recursion. This function adds up the elements of its list; for example, `sum [1,2,3]` returns $1+2+3 = 6$. The type of `sum` reflects the fact that the elements of a list must be numbers if you want to add them up. The type context 'Num a =>' says that a can stand for any type provided that the numeric operations are defined. Thus `sum` could be applied to a list of 32-bit integers, or a list of unbounded integers, or a list of floating point numbers, or a list of complex numbers, and so on.

The definition has the same form as the previous example. We define the sum of an empty list to be 0; this is required to make the recursion work, and it makes sense anyway. It is common to define the base case of functions so that recursions work properly. This also usually gives good algebraic properties to the function. (This is the reason that 0 ! is defined to be 1.) For the recursive case, we add the head of the list x to the sum of the rest of the elements, computed by the recursive call `sum xs`.

```
sum :: Num a => [a] -> a
sum [] = 0
sum (x:xs) = x + sum xs
```

The value of sum [1,2,3] can be calculated using equational reasoning:

```
sum [1,2,3]
= 1 + sum [2,3]
= 1 + (2 + sum [3])
= 1 + (2 + (3 + sum []))
= 1 + (2 + (3 + 0))
= 6
```

So far, the functions we have written receive a list and return a number. Now we consider functions that return lists. A typical example is the (++) function (its name is often pronounced either as 'append' or as 'plus plus'). This function takes two lists and appends them together into one bigger list. For example, [1,2,3] ++ [9,8,7,6] returns [1,2,3,9,8,7,6].

Notice that this function has *two* list arguments. The definition uses recursion over the first argument. It's easy to figure out the value of [] ++ [a,b,c]; the first list contributes nothing, so the result is simply [a,b,c]. This observation provides a base case, which is essential to make recursion work: we can define [] ++ ys = ys. For the recursive case we have to consider an expression of the form (x:xs) ++ ys. The first element of the result must be the value x. Then comes a list consisting of all the elements of xs, followed by all the elements of ys; that list is simply xs++ys. This suggests the following definition:

```
(++) :: [a] -> [a] -> [a]
[] ++ ys = ys
(x:xs) ++ ys = x : (xs ++ ys)
```

Working out an example is a good way to check your understanding of the definition:

```
[1,2,3] ++ [9,8,7,6]
= 1 : ([2,3] ++ [9,8,7,6])
= 1 : (2 : ([3] ++ [9,8,7,6]))
= 1 : (2 : (3 : ([] ++ [9,8,7,6])))
= 1 : (2 : (3 : [9,8,7,6]))
= 1 : (2 : [3,9,8,7,6])
= 1 : [2,3,9,8,7,6]
= [1,2,3,9,8,7,6]
```

Once we know the structure of the definition—a recursion over xs—it is straightforward to work out the equations. The trickiest aspect of writing the definition of (++) is deciding over which list to perform the recursion. Notice that the definition just treats ys as an ordinary value, and never checks whether it is empty or non-empty. But you can't always assume that if there

are several list arguments, the recursion will go over the first one. Some functions perform recursion over the second argument but not the first, and some
functions perform a recursion simultaneously over *several* list arguments.

The `zip` function is a typical example of a function that takes two list arguments and performs a recursion simultaneously over both of them. This function takes two lists, and returns a list of pairs of elements. Within a pair, the
first value comes from the first list, and the second value comes from the second
list. For example, `zip [1,2,3,4] ['A', '*', 'q', 'x']` returns `[(1,'A'),
(2,'*'), (3,'q'), (4,'x')]`. There is a special point to watch out for: the
two argument lists might have different lengths. In this case, the result will
have the same length as the shorter argument. For example, `zip [1,2,3,4]
['A', '*', 'q']` returns just `[(1,'A'), (2,'*'), (3,'q')]` because there
isn't anything to pair up with the 4.

The definition of `zip` must do a recursion over *both* of the argument lists,
because the two lists have to stay synchronised with each other. There are two
base cases, because it's possible for either of the argument lists to be empty.
There is no need to write a third base case `zip [] [] = []`, since the first
base case will handle that situation as well.

```
zip :: [a] -> [b] -> [(a,b)]
zip [] ys = []
zip xs [] = []
zip (x:xs) (y:ys) = (x,y) : zip xs ys
```

Here is a calculation of the example above. The recursion terminates when the
second list becomes empty; the second base case equation defines the result of
`zip [4] []` to be `[]` even though the first argument is non-empty.

```
  zip [1,2,3,4] ['A', '*', 'q']
= (1,'A') : zip [2,3,4] ['*', 'q']
= (1,'A') : ((2,'*') : zip [3,4] ['q'])
= (1,'A') : ((2,'*') : ((3,'q') : zip [4] []))
= (1,'A') : ((2,'*') : ((3,'q') : []))
= (1,'A') : ((2,'*') : [(3,'q')])
= (1,'A') : [(2,'*'), (3,'q')]
= [(1,'A'), (2,'*'), (3,'q')]
```

The `concat` function takes a list of lists and flattens it into a list of elements.
For example, `concat [[1], [2,3], [4,5,6]]` returns one list consisting of all
the elements in the argument lists; thus the result is `[1,2,3,4,5,6]`.

```
concat :: [[a]] -> [a]
concat [] = []
concat (xs:xss) = xs ++ concat xss
```

In working out the example calculation, we just simplify all the applications
of ++ directly. Of course each of those applications entails another recursion,
which is similar to the examples of (++) given above.

```
concat [[1], [2,3], [4,5,6]]
= [1] ++ concat [[2,3], [4,5,6]]
= [1] ++ ([2,3] ++ concat [[4,5,6]])
= [1] ++ ([2,3] ++ [4,5,6])
= [1] ++ [2,3,4,5,6]
= [1,2,3,4,5,6]
```

The base case for a function that builds a list must return a list, and this is often simply []. The recursive case builds a list by attaching a value onto the result returned by the recursive call.

In defining a recursive function f, it is important for the recursion to work on a list which is shorter than the original argument to f. For example, in the application sum [1,2,3], the recursion calculates sum [2,3], whose argument is one element shorter than the original argument [1,2,3]. If a function were defined incorrectly, with a recursion that is bigger than the original problem, then it could just go into an infinite loop.

All of the examples we have seen so far perform a recursion on a list which is one element shorter than the argument. However, provided that the recursion is solving a smaller problem than the original one, the recursive case can be anything—it doesn't necessarily have to work on the tail of the original list. Often a good approach is to try to cut the problem size in half, rather than reducing it by one. This organisation often leads to highly efficient algorithms, so it appears frequently in books on the design and analysis of algorithms. We will look at the quicksort function, a good example of this technique.

Quicksort is a fast recursive sorting algorithm. The base case is simple: quicksort [] = []. For a non-empty list of the form (x:xs), we will first pick one element called the *splitter.* For convenience that will be x, the first element in the list. We will then take all the elements of the rest of the list, xs, which are *less than or equal to* the splitter. We will call this list the *small elements,* and define it as a list comprehension [y | y <- xs, y<=splitter]. In a similar way we define the list of *large elements,* which are greater than the splitter, as [y | y <- xs, y>splitter]. Now the complete sorted list consists first of the small elements (in sorted order), followed by the splitter, followed by the large elements (in sorted order).

```
quicksort :: Ord a => [a] -> [a]
quicksort [] = []
quicksort (splitter:xs) =
  quicksort [y | y <- xs, y<=splitter]
  ++ [splitter]
  ++ quicksort [y | y <- xs, y>splitter]
```

It is interesting to compare this definition with a conventional one in an imperative language; see any standard book on algorithms.

Exercise 1. Write a recursive function copy :: [a] -> [a] that copies its list argument. For example, copy [2] ⇒[2].

Exercise 2. Write a function `inverse` that takes a list of pairs and swaps the pair elements. For example,

```
inverse [(1,2),(3,4)] ==> [(2,1),(4,3)]
```

Exercise 3. Write a function

```
merge :: Ord a => [a] -> [a] -> [a]
```

which takes two sorted lists and returns a sorted list containing the elements of each.

Exercise 4. Write (!!), a function that takes a natural number n and a list and selects the *n*th element of the list. List elements are indexed from 0, not 1, and since the type of the incoming number does not prevent it from being out of range, the result should be a `Maybe` type. For example,

```
[1,2,3]!!0 ==> Just 1
[1,2,3]!!2 ==> Just 3
[1,2,3]!!5 ==> Nothing
```

Exercise 5. Write a function `lookup` that takes a value and a list of pairs, and returns the second element of the pair that has the value as its first element. For example,

```
lookup 5 [(1,2),(5,3)] ==> Just 3
```

Exercise 6. Write a function that counts the number of times an element appears in a list.

Exercise 7. Write a function that takes a value e and a list of values xs and removes all occurrences of e from xs.

Exercise 8. Write a function

```
f :: [a] -> [a]
```

that removes alternating elements of its list argument, starting with the first one. For examples, f [1,2,3,4,5,6,7] returns [2,4,6].

Exercise 9. Write a function that takes a list of `Maybe` values and returns the elements they contain.

Exercise 10. Write a function

```
f :: String -> String -> Maybe Int
```

that takes two strings. If the second string appears within the first, it returns the index identifying where it starts. Indexes start from 0. For example,

```
f "abcde" "bc" ==> Just 1
f "abcde" "fg" ==> Nothing
```

5.2 Higher Order Recursive Functions

Many of the recursive definitions from the previous section are quite similar to
each other. An elegant idea is to write a function that expresses the general
computation pattern once and for all. This general function could then be
reused to define a large number of more specific functions, without needing
to write out the complete recursive definitions. In order to allow the general
function to know exactly what computation to perform, we need to supply it
with an extra argument which is itself a function. Such a general function is
called a *higher order function* or a *combinator*.

Several of the functions defined in the previous section take a list argument,
and return a list result, where each element of the result is computed from the
corresponding element of the input. We can write a general function, called
map, which expresses all particular functions of that form. The first argument
to map is another function, of type a->b, which takes a single value of type
a and returns a single result of type b. The purpose of map is to apply that
auxiliary function to all the elements of the list argument.

```
map :: (a->b) -> [a] -> [b]
map f []    = []
map f (x:xs) = f x : map f xs
```

Here is an example of the use of map. Suppose the function we define is
to multiply every list element by 5. We could write a special function to do
just that, using recursion. A simpler method is to specify the 'multiply by 5'
function (*5) as an argument to map. Here is an example:

```
map (*5) [1,2,3]
= 1*5 : map (*5) [2,3]
= 1*5 : (2*5 : map (*5) [3])
= 1*5 : (2*5 : (3*5 : map (*5) []))
= 1*5 : (2*5 : (3*5 : []))
= 5 : (10 : (15 : []))
= [5,10,15]
```

The map function takes an auxiliary function requiring one argument, and
one list argument. Sometimes we have an auxiliary function that takes two
arguments, and want to apply it to all the corresponding elements of two lists.
If the argument lists are of different sizes, the result will have the length of
the shorter one, This is performed by the zipWith function, which performs a
recursion simultaneously over both list arguments:

```
zipWith :: (a->b->c) -> [a] -> [b] -> [c]
zipWith f [] ys = []
zipWith f xs [] = []
zipWith f (x:xs) (y:ys) = f x y : zipWith f xs ys
```

Some of the recursive functions from the previous section, such as `length` and `sum`, have a structure which is slightly different from that of `map`. These functions take a list but return a singleton result which is calculated by combining elements from the list. A general function for this is `foldr`, which is defined as follows:

```
foldr :: (a->b->b) -> b -> [a] -> b
foldr f z [] = z
foldr f z (x:xs) = f x (foldr f z xs)
```

Here is an example where `foldr` is used to add up the elements of a list. The singleton argument `z` can be thought of as an initial value for the sum, so we specify 0 for its value.

```
foldr (+) 0 [1,2,3]
= 1 + foldr (+) 0 [2,3]
= 1 + (2 + foldr (+) 0 [3])
= 1 + (2 + (3 + foldr (+) 0 []))
= 1 + (2 + (3 + 0))
= 6
```

The function's name suggests that a list is folded into a singleton value. The recursion produces a sequence of applications of the `f` function, starting from the right end of the list (hence the name, which stands for *fold from the right*). The singleton argument `z` serves to initialise the accumulator, and it also provides a result in case the entire list argument is empty.

Now we can define all the functions which follow this general pattern with a single equation; there is no need to keep writing out separate recursive definitions. Here is a collection of useful functions, all definable with `foldr`:

```
sum xs        = foldr (+) 0 xs
product xs    = foldr (*) 1 xs
and xs        = foldr (&&) True xs
or xs         = foldr (||) False xs
factorial n   = foldr (*) 1 [1..n]
```

We now have two definitions of sum, a recursive one and one using `foldr`. These two definitions ought to produce the same result, and it is interesting to consider in more detail how this comes about. The recursive definition, which we saw in the previous section, is:

```
sum [] = 0
sum (x:xs) = x + sum xs
```

The definition using `foldr` is:

```
sum xs        = foldr (+) 0 xs
```

First, consider `sum []`. The recursive definition says directly that this has the value 0. We can calculate what the `foldr` definition says:

```
sum []
= foldr (+) 0 []
= 0
```

The two definitions give the same result for the base case, as they should. Now consider the recursive case, where the argument has the structure `(x:xs)`. The recursive definition returns `x + sum xs`. The result produced by the definition with `foldr` is calculated:

```
sum (x:xs)
= foldr (+) 0 (x:xs)
= x + foldr (+) 0 xs
= x + sum xs
```

We have just proved that the two definitions of `sum` always produce the same result, since *every* list must either be `[]` or have the form `(x:xs)`.

Often there is a choice between using a recursive definition, or using one of the existing higher order functions like `map`, `foldr` etc. For example, consider the problem of writing a function `firsts`, which should take a list of pairs and returns a list of the first elements. For example,

```
firsts [(1,3),(2,4)] ==> [1,2]
```

Here is a recursive definition:

```
firsts :: [(a,b)] -> [a]
firsts [] = []
firsts ((a,b):ps) = a : firsts ps
```

And here is an alternative definition using `map`:

```
firsts :: [(a,b)] -> [a]
firsts xs = map fst xs
```

The definition using `map` is generally considered better style. It is shorter and more readable, but its real advantage is modularity: the definition of `map` expresses a specific form of recursion, and the definition of `firsts` with `map` makes it explicitly clear that this form of recursion is being used.

Exercise 11. Write `foldrWith`, a function that behaves like `foldr` except that it takes a function of three arguments and two lists.

Exercise 12. Using `foldr`, write a function `mappend` such that

```
mappend f xs = concat (map f xs)
```

Exercise 13. Write `removeDuplicates`, a function that takes a list and removes all of its duplicate elements.

Exercise 14. Write a recursive function that takes a value and a list of values and returns True if the value is in the list and False otherwise.

5.3 Recursion Over Trees

Recursion really comes into its own when more complex data structures are used, such as trees or graphs. The recursions we used over lists may not obviously be better than ordinary iterations, such as **for** or **while** loops. However, recursion is almost always far superior to ordinary loops for trees. The use of recursion is not restricted to functional languages; you will also use it for working with complex data structures in imperative and object-oriented languages.

There are many different ways to define trees. For this reason there is no single standard built-in tree data structure in Haskell. In contrast there is just one sensible way to define lists, so you don't have to define them yourself; the list type is built-in. The particular definition we give below, and the functions that operate on it, are flexible and commonly used, but it is also straightforward to modify these definitions in order to tailor the trees for a particular application.

A tree (as defined in this section) has type `Tree a`, where `a` is the type of data values that may be stored in the tree nodes. A tree may be one of two things:

- A `Tip` contains no information, and is analogous to the empty list `[]`;

- A `Node` contains a piece of data with type `a`, and two subtrees of type `Tree a`.

The data structure is defined as an algebraic data type, with the constructors `Tip` and `Node`. The clause at the end, `deriving Show`, tells the Haskell compiler to produce code automatically that can convert a tree to a character string for printing; you don't have to do this yourself.

```
data Tree a
  = Tip
  | Node a (Tree a) (Tree a)
  deriving Show
```

We can define a tree simply using the constructors `Tip` and `Node` as functions that construct a particular tree (that is why they are called constructors). Here are some examples, and Figure 5.1 shows their corresponding diagrams.

```
t1, t2 :: Tree Int
t1 = Node 6 Tip Tip
t2 = Node 5
        (Node 3 Tip Tip)
        (Node 8 (Node 6 Tip Tip) (Node 12 Tip Tip))
```

Now we consider several functions that perform operations on trees, and which are defined using recursion. The `nodeCount` function takes a tree and determines how many nodes are in it; this is also the number of data values stored in the tree. The base case is an empty tree, a `Tip`, where the number of

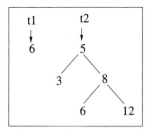

Figure 5.1: Tree Diagrams

nodes is 0. The recursive case is a tree consisting of a `Node`, with a data value
x and two subtrees `t1` and `t2`. There are two recursive calls, since there are
two trees inside the `Node`.

```
nodeCount :: Tree a -> Int
nodeCount Tip = 0
nodeCount (Node x t1 t2) = 1 + nodeCount t1 + nodeCount t2
```

The next function reflects a tree around a vertical axis; the left and right
subtrees are interchanged and the same reflection happens recursively within
the subtrees.

```
reflect :: Tree a -> Tree a
reflect Tip = Tip
reflect (Node a t1 t2) = Node a (reflect t2) (reflect t1)
```

The `mapTree` function is analogous to `map`, which operates over lists.

```
mapTree :: (a->b) -> Tree a -> Tree b
mapTree f Tip = Tip
mapTree f (Node a t1 t2) =
   Node (f a) (mapTree f t1) (mapTree f t2)
```

Trees are commonly used in efficient algorithms that perform searching
quickly. In a *binary search tree*, all the data values that are smaller than the
data in the node are stored in the left subtree, while the right subtree holds all
the larger values.

The binary search tree is efficient because, if it is balanced (i.e. its left and
right subtrees have the same height) then every step in the recursion cuts out
half of the remaining possible data values. A binary search zeros in quickly on
the desired value, instead of creeping toward it one value at a time as with list
searches.

A common technique is to store a (`key,value`) pair in each node, and
to use the `key` fields to organise the structure of the binary search tree; for
example:

```
tree :: Tree (Int,Int)
tree =
  Node (5,10)
    (Node (3,6)   (Node (1,1) Tip Tip)
                  (Node (4,8) Tip Tip))
    (Node (7,14)  (Node (6,12) Tip Tip)
                  (Node (8,16) Tip Tip))
```

Now, suppose that we want to find the node that contains the key 6, so that the corresponding value, 12, can be obtained. The search function will compare the desired key 6 with the root node's key; since that is 5 it the search will consider only the right subtree, inspect 7 then choose its left subtree. The find function, implementing a binary search over a tree, is defined as follows:

```
find :: Int -> Tree (Int,a) -> Maybe a
find n Tip = Nothing
find n (Node (m,d) t1 t2) =
  if n==m then Just d
  else if n<m then find n t1
              else find n t2
```

Exercise 15. Write appendTree, a function that takes a tree and a list and appends the contents of the tree (traversed from left to right) to the front of the list. For example,

```
appendTree (Node 1 (Node 2 Tip Tip)
                   (Node 3 Tip Tip))
           [4,5]
  ==> [2,1,3,4,5]
```

Exercise 16. Write concatTree, a function that takes a tree of lists and concatenates the lists in order from left to right. For example,

```
concatTree (Node [2] (Node [3,4] Tip Tip)
                     (Node [5] Tip Tip))
  ==> [3,4,2,5]
```

Exercise 17. Write zipTree, a function that takes two trees and pairs each of the corresponding elements in a list. Your function should return Nothing if the two trees do not have the same shape. For example,

```
zipTree (Node 2 (Node 1 Tip Tip) (Node 3 Tip Tip))
        (Node 5 (Node 4 Tip Tip) (Node 6 Tip Tip))
  ==> Just [(1,4),(2,5),(3,6)]
```

Exercise 18. Write zipWithTree, a function that is like zipWith except that it takes trees instead of lists. It returns a list. You can assume that the two arguments will have the same shape.

5.4   Peano Arithmetic

We turn now to the implementation of arithmetic operations over a simple data structure representing the natural numbers. The reason for working with this representation is that it provides a good introduction to recursion over general algebraic data types.

```
data Peano = Zero | Succ Peano deriving Show
```

For example,

```
1 = Succ Zero
3 = Succ (Succ (Succ Zero))
```

As you can see, the Peano data type is recursive. In this case, the recursion builds up a series of constructor applications, somewhat like the list data type:

```
data List a = Empty | Cons a (List a)

[1]     = Cons 1 Empty
[1,2,3] = Cons 1 (Cons 2 (Cons 3 Empty))
```

The simplest Peano function is decrement, which removes a Succ constructor if possible, returning Zero otherwise:

```
decrement :: Peano -> Peano
decrement Zero     = Zero
decrement (Succ a) = a
```

The definition of add is:

```
add :: Peano -> Peano -> Peano
add Zero     b = b
add (Succ a) b = Succ (add a b)
```

This definition looks a lot like that of (++)!

However, sub is more complex. If the second argument is Zero, then sub returns the first argument. Negative numbers cannot be represented in this scheme, so they are approximated by Zero. Otherwise sub must decrement both numbers, as follows:

```
sub :: Peano -> Peano -> Peano
sub a        Zero     = a
sub Zero     b        = Zero
sub (Succ a) (Succ b) = sub a b
```

We can do more with recursion and Peano numbers. Here are two predicates, equals and lt:

```
equals :: Peano -> Peano -> Bool
equals Zero      Zero        = True
equals Zero      b           = False
equals a         Zero        = False
equals (Succ a) (Succ b)     = equals a b


lt :: Peano -> Peano -> Bool
lt a         Zero       = False
lt Zero      (Succ b)   = True
lt (Succ a) (Succ b)    = lt a b
```

5.5 Data Recursion

So far we have always used recursion to define functions. Recursive functions are useful in nearly all programming languages, and they are especially important for programming with data structures like trees and graphs. There is another important programming technique based on recursion, however, which can be used only in *nonstrict* programming languages such as Haskell. This technique is called *data recursion.*

The idea is to define *circular* data structures. Here is one way to define an infinitely long list where every element is 1:

```
f :: a -> [a]
f x = x : f x
ones = f 1
```

Each time the recursive function f is applied, it uses (:) to construct a new list element containing x. As a result, there is no bound on how much computer memory will be required to represent ones; this depends on how much of the computation is actually demanded by the rest of the program.

However, it's possible to represent ones very compactly with a circular list, defined with recursion in the data rather than in a function:

```
twos = 2 : twos
```

Figure 5.2 shows the internal representations of ones and twos. Mathematically, the *value* of both is a list which is infinitely long. However, the circular definition requires far less memory.

A data structure consisting of nodes connected by links is called a *graph.* Graphs can be constructed by defining each node with an equation in a let expression; this means that each node can be referred to by any other node (including even itself). For example, the following Haskell definition creates a circular data structure, where the internal names a, b and c are used to set up the links:

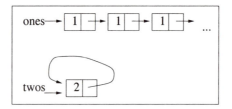

Figure 5.2: Circular and Non-circular Infinite Lists

```
object = let a = 1:b
             b = 2:c
             c = [3] ++ a
         in a
```

5.6   Suggestions for Further Reading

Most programming languages have a variety of control constructs for looping, but recursion is general enough to encompass all of them. The textbooks on Haskell (see Chapter 1) give many examples, and Abelson and Sussman [1] use recursion to provide a variety of control abstractions in the programming language Scheme.

Systems with self-reference, including especially recursion, are one of the major themes of Hofstadter's book *Gödel, Escher, Bach* [16].

Recursion is central to the theory of computation; one of the main theoretical subjects in computer science is called *recursive function theory*. This is one of the standard topics in computer science, and many textbooks are available.

5.7   Review Exercises

Exercise 19. Write a function that uses recursion to take the intersection of two sets.

Exercise 20. Using recursion, write a function that takes two sets and returns True if the first is a subset of the other.

Exercise 21. Write a recursive function that determines whether a list is sorted.

Exercise 22. Show that the definition of factorial using foldr always produces the same result as the recursive definition given in the previous section.

Exercise 23. Using recursion, define last, a function that takes a list and returns a Maybe type that is Nothing if the list is empty.

Exercise 24. Using recursion, write two functions that expect a string containing a real number. The first returns the whole part of the number, to the left of the decimal point, and the second returns the fractional part to the right of the decimal point.

Chapter 6

Inductively Defined Sets

In this chapter, we explore the construction of sets using induction. To understand why induction is useful, consider the problem of defining a set. The simplest method is to define a set by naming each of its elements, one by one. This is called *enumeration*. It works only for finite sets and is impractical for large sets. Another approach is to use ellipses ('...') to indicate that the set continues, but this is imprecise, and so can be ambiguous. For example, what is meant by $\{1, 2, 3, \ldots\}$? Is the next element 4 or 5? Even if you think the answer is obvious, how do you know everyone else will consider the *same* thing to be obvious? If we are enumerating the positive integers, the next element is 4, but if we are adding the two previous numbers in the series, it is 5.

6.1 The Idea Behind Induction

Induction is rather like a mathematical 'program' that calculates a proof when needed. The proof asserts that an element is a member of the set defined by induction. For example, here are two propositions:

$$0 \in S$$
$$n \in S \rightarrow n + 1 \in S$$

Together, they let us show that any natural number is in set S. To see how they do this, consider an example: we show that 2 is an element of set S. Using the propositions above, we can construct a *chain* that looks like this:

$$0 \in S$$
$$0 \in S \rightarrow 1 \in S$$
$$1 \in S \rightarrow 2 \in S$$

We then use the first assertion and *Modus Ponens* to deduce that 1 is in set S, and from that, using *Modus Ponens*, we deduce that 2 is in set S. In fact, we can use the two propositions to build a chain that is as long as needed to reach *any* natural number.

Of course, if any of the links in the chain were missing, for example the proposition $0 \in S \to 1 \in S$, then we could not reach the number required. This is because we could not use *Modus Ponens* to get to the next link.

When we use an inductive definition to show that a set contains a given value v, we enumerate, or count, the values that must first be shown to be in the set before v. These values form a *sequence*, which is a set with an ordering. Computers can enumerate elements of sets in the order in which they are generated from a description of the set. We can use a computer to calculate a sequence that represents an infinite set, although we will only see a finite prefix of the entire sequence.

Let's implement what we have seen using Haskell. A set of numbers can be represented by a list; for example, the set with the numbers 1,2 and 3 is [1,2,3], and the empty set is [].

How do we implement the implications? An implication of the form $1 \in s \to 2 \in s$ can be implemented as a function that takes 1 and returns the next element, 2. If it is applied to anything other than 1, then an error message is returned:

```
imp1 :: Int -> Int
imp1 1 = 2
imp1 other = error "premise does not match"
```

We can implement a chain using function application. The argument of the function imp1 is an element of s. If that element matches the pattern of imp1, then imp1 can be applied to it and produce a new element of s. This is just like what we do when deciding whether we can use *Modus Ponens*: we match the premise of the implication with elements of the set; if there is a match, then we can use *modus ponens*, otherwise the match fails and we cannot. For example, consider the following assertions:

$$1 \in S$$
$$1 \in S \to 2 \in S$$
$$2 \in S \to 3 \in S$$

This can be implemented by the following Haskell definitions:

```
imp1 :: Int -> Int
imp1 1 = 2
imp1 x = error  "premise does not match"

imp2 :: Int -> Int
imp2 2 = 3
imp2 x = error  "premise does not match"
```

```
s :: [Int]
s = [1, imp1 (s !! 0), imp2 (s !! 1)]
```

The function application s!!0 returns the first element of s, indexing from 0; the result is 1. The function imp11 is applied to this. Since the argument matches the pattern, the application succeeds, adding 2 to s. Then the function imp12 is applied to 2. Since the argument matches the pattern, the application succeeds, adding 3 to s. The value of s is [1,2,3].

There is a difference between matching the premise of the implication with all of the elements of the set and applying imp1 to the most recently added member of s. However, in the form of induction that we are studying, it doesn't matter. The only value that could possibly match the premise is the one generated by the previous implication. But consider now this set of assertions:

$$1 \in S$$
$$2 \in S \rightarrow 3 \in S$$

This is implemented as:

```
imp1 :: Int -> Int
imp1 2 = 3
imp1 x = error   "premise does not match"

s :: [Int]
s = [1, imp1 (s !! 0)]
```

In this case, we cannot use *modus ponens* to conclude that 3 is in the list, because nothing states that 2 is in it.

Exercise 1. Is the following a chain? You can test your conclusions by evaluating s in each case.

```
imp1 :: Int -> Int
imp1 1 = 2
imp1 x = error "imp1: premise does not apply"

imp2 :: Int -> Int
imp2 2 = 3
imp2 x = error "imp2: premise does not apply"

imp3 :: Int -> Int
imp3 3 = 4
imp3 x = error "imp3: premise does not apply"
s :: [Int]
s = [1, imp1 (s !! 0), imp2 (s !! 1), imp3 (s !! 2)]
```

Exercise 2. Is the following a chain?

```
imp1 :: Int -> Int
imp1 1 = 2
imp1 x = error "imp1: premise does not apply"

imp2 :: Int -> Int
imp2 3 = 4
imp2 x = error "imp2: premise does not apply"

s :: [Int]
s = [0, imp1 (s !! 0), imp2 (s !! 1)]
```

Exercise 3. Is the following a chain?

```
imp1 :: Int -> Int
imp1 0 = 1
imp1 x = error "imp1: premise does not apply"

imp2 :: Int -> Int
imp2 3 = 4
imp2 x = error "imp2: premise does not apply"

s :: [Int]
s = [0, imp1 (s !! 0), imp2 (s !! 1)]
```

Exercise 4. Is the following a chain?

```
imp1 :: Int -> Int
imp1 0 = 1
imp1 x = error "imp1: premise does not apply"

imp2 :: Int -> Int
imp2 1 = 2
imp2 x = error "imp2: premise does not apply"

s :: [Int]
s = [0, imp1 (s !! 1), imp2 (s !! 0)]
```

6.1.1   The Induction Rule

Recall the two propositions we used in the first section:

$$0 \in S$$
$$n \in S \rightarrow n + 1 \in S$$

The first one is called the *base case*, and the second is called the *induction case*, or the *induction rule.*. It is the induction case that generates the links of the chain which will allow us to reach any number in the set being defined.

So far, the induction has had a fixed form because we were defining a particular set. However, we could have a rule

$$n \in S \rightarrow n + 2 \in S.$$

Together with the base case $1 \in S$, this would define the odd natural numbers. Alternatively, our induction rule might be

$$n \in S \rightarrow n * 5 \in S.$$

Together with the base case, this would define the set of powers of 5.

As we will see, it can sometimes be hard to construct the correct rule, and it is necessary to debug rules to get them right.

Suppose we have a set defined by the following assertions:

$$0 \in s$$
$$x \in s \rightarrow x + 1 \in s$$

First, we want to find out whether 2 is in the set, and will use the computer to help. (Of course, we could solve this by hand, but if the induction rule is complicated, or if we want to find out whether 1,000,000 is in the set, software tools are invaluable.) We can implement the induction rule as the `increment` function:

```
increment :: Int -> Int
increment x = x + 1

s :: [Int]
s = [0, increment (s !! 0), increment (s !! 1)]
```

We can load this definition and evaluate s; the last element of s is 2.

Now suppose that we want to know whether 50 is in the set. It would be very tedious to write out each element as we have been doing. The same function is applied to each element of s, so we can have the following definition of s instead:

```
s :: [Int]
s = 0 : map increment s
```

The function `map` proceeds down s, creating each value it needs and then using it. We can then get at the fiftieth element by typing s!!50.

Now we have a new format for implementing inductive definitions. We first specify the induction rule, then recursively define a list in which the base case appears first, and then the rule is mapped down the list.

Exercise 5. Given the base case $0 \in n$ and the induction rule $x \in n \to x+1 \in$ n, fix the following calculation so that 3 is in set n:

```
fun :: Int -> Int
fun x = x - 1

n :: [Int]
n = 0 : map fun n
```

Exercise 6. Use the following definitions, determine whether 4 is in set s, given $1 \in s$ and the induction rule $x \in s \to x + 2 \in s$.

```
fun :: Int -> Int
fun x = x + 2

s :: [Int]
s = 1 : map fun s
```

Exercise 7. Fix this calculation of the positive integers:

```
fun :: Int -> Int
fun x = 0

p :: [Int]
p = 0 : map fun p
```

Exercise 8. Fix this calculation of the positive multiples of 3:

```
fun :: Int -> Int
fun x = x * 3

p :: [Int]
p = map fun p
```

6.2   How to Define a Set Using Induction

We have seen that an inductive set definition has a base case and an induction rule (or induction case). There is one more clause that needs to be specified in an inductive set definition. Suppose that we have defined a set S by saying that the numbers 1, 2 and 3 are in S. How do we know that *something else* isn't also in S? If we don't say explicitly that nothing else is in S, then the specification could be satisfied by lots of different sets. It could be the set $\{1, 2, 4.5, -78, 3\}$, for example.

We want to exclude all elements that aren't introduced by the base case, or instantiations of the induction case, so we include a clause (called the extremal clause) in a set definition that states *Nothing is an element of the set unless it can be constructed by a finite number of uses of the first two clauses.*

To summarise, an inductive definition of a set consists of three parts: a base case, an induction case and an extremal clause:

- The base case is a simple statement of some mathematical fact, such as $1 \in$ S;

- The induction case is an implication in a general form, such as the proposition that
$$\forall x : U, x \in S \rightarrow x + 1 \in S.$$

- The extremal clause says that nothing is in the set being defined unless it got there by a finite number of uses of the first two cases.

6.2.1   Inductive Definition of the Set of Natural Numbers

We will illustrate the method by writing an inductive definition of the natural numbers.

Definition 19. The set N of natural numbers is defined as follows:

- Base case: $0 \in N$

- Induction case: $x \in N \rightarrow x + 1 \in N$

- Extremal clause: nothing is an element of the set N unless it can be constructed with a finite number of uses of the base and induction cases.

Now we can use the base and induction cases to show formally that an arbitrary number above and including 0 is a natural number. Let's choose the number 2.

1.	$0 \in N$	Base case
2.	$0 \in N \rightarrow 1 \in N$	instantiation rule, induction case
3.	$1 \in N$	1, 2, *Modus Ponens*
4.	$1 \in N \rightarrow 2 \in N$	instantiation rule, induction case
5.	$2 \in N$	3, 4, *Modus Ponens*

Exercise 9. Here is a Haskell equation that defines the set s inductively. Is 82 an element of s?

```
s :: [Int]
s = 0 : map ((+) 2) s
```

Exercise 10. What set is defined by the following?

```
s :: [Int]
s = 1 : map ((*) 3) s
```

Exercise 11. Do we need to implement the extremal clause in a Haskell specification? If so, how can this be done? If not, why not?

6.2.2 The Set of Binary Machine Words

Now we define a set `BinWords`, each of which is a machine word represented in binary notation. In general, a machine word can be of any length.

The base case says that the elements of another set (the set of binary digits) are also elements of `BinWords`. The induction case uses *concatenation* to create a new value from one already in the set. We represent the concatenation of a character to a string by placing them one after the other: e.g. '1' '01' is the string '101'.

Definition 20. Let `BinDigit` be the set $\{0, 1\}$. The set `BinWords` of machine words in binary is defined as follows:

- Base case:

$$x \in \text{BinDigit} \;\rightarrow\; x \in \text{BinWords}$$

- Induction case: if x is a binary digit and y is a binary word, then their concatenation xy is also a binary word:

$$(x \in \text{BinDigit} \wedge y \in \text{BinWords}) \;\rightarrow\; x\,y \in \text{BinWords}$$

- Extremal clause: nothing is an element of `BinWords` unless it can be constructed with a finite number of uses of the base and induction cases.

A set based on another set S in this way is given the name S^{+}, indicating that it is the set of all possible non-empty strings over S. The expression S^{*} is the same as S^{+} except that S^{*} also includes the empty string. Thus our set `BinWords` could be written BinDigit^{+}.

We can write a Haskell function to calculate the set of binary words, using the inductive definition just presented. The induction function takes a binary word. It creates a new one from that number and each binary digit in turn. For example, if it is given `[1,0]`, it returns `[0,1,0]` and `[1,1,0]`.

The induction rule takes a binary word and creates two new ones, so we define the function `newBinaryWords` to do just that:

```
newBinaryWords :: [Int] -> [[Int]]
newBinaryWords ys = [0 : ys, 1 : ys]
```

Finally, we define the set of binary words as follows:

```
mappend :: (a -> [b]) -> [a] -> [b]
mappend f []     = []
mappend f (x:xs) = f x ++ mappend f xs

binWords = [0] : [1] :
                (mappend newBinaryWords binWords)
```

Exercise 12. Alter the definition of `newBinaryWords` and `binWords` so that they produce all of the *octal* numbers. An octal number is one that contains only the digits 0 through 7.

6.3 Defining the Set of Integers

Now we come to a more subtle problem: defining the set I of integers. Instead of just giving the final answer, we will think through the problem the way you might in real life. Each time we have a trial solution, we will examine it to see if it works, and whether it can be improved.

The sets we have been defining are *well-founded*; that is, they are infinite only in one direction, and they have a least element. The set N of natural numbers is a good example of a well-founded set; in fact, a *countable* set is one that can be counted using the natural numbers (see Chapter 9).

Are the integers countable? They don't have a least element, and they are infinite in two directions, so they aren't well-founded. But we could use a trick: start from 0 and count first n then $-n$. If we think of the integers as a measuring tape that is infinitely long in two directions, we could still count the inches (or centimetres) on the tape by thinking of the tape as being folded at 0. Now the positive numbers touch the negative numbers, and each element i of the naturals counts both the positive and the negative numbers; that is, i counts $(i, -i)$.

We have just devised a way of enumerating the integers so that every element is eventually counted. This forms an excellent basis for an inductive definition of the integers. We will now work through several attempts to define the integers using induction. Some of these have problems, and we discuss how to debug them.

6.3.1 First Attempt

Attempt 1. The set I is defined as follows:

- Base case: $0 \in I$

- Induction case: $x \in I \rightarrow -x \in I$

- Extremal clause: nothing is in I unless its presence is justified by a finite number of uses of the base and induction cases.

We will now define some functions that will make our inductive definitions easier to understand, and also make it possible to use the computer to help carry out experiments with the definition. Each of these is designed to take the base and induction cases as arguments and construct the data recursion automatically. The first uses map and the second uses mappend.

```
build :: a -> (a -> a) -> Set a
build a f = set
            where set = a : map f set

builds :: a -> (a -> [a]) -> Set a
builds a f = set
```

```
where set = a : mappend f set
```

Here is an implementation of the first definition:

```
nextInteger1 :: Int -> Int
nextInteger1 x = -x

integers1 :: [Int]
integers1 = build 0 nextInteger1
```

Exercise 13. Use `take 10 integers1` to evaluate the first 10 integers according to this definition. Describe the set which is actually defined by Attempt 1.

6.3.2   Second Attempt

Attempt 1 doesn't work. It was based on our intuitive method making the integers countable. The problem is that according to this definition, only 0 is a member of I. When the natural numbers were defined, there was a mechanism for including new numbers by adding 1 to the integer in the premise. Something similar is needed here to include the other integers.

Attempt 2. The set I is defined as follows:

- Base case: $0 \in I$

- Induction case: $x \in I \rightarrow (x + 1 \in I \wedge x - 1 \in I)$

- Extremal clause: Nothing is in I unless its presence is justified by a finite number of uses of the base and induction cases.

Here is an implementation of Attempt 2:

```
nextIntegers2 :: Int -> [Int]
nextIntegers2 x = [x + 1, x - 1]

integers2 :: [Int]
integers2 = builds 0 nextIntegers2
```

Exercise 14. Use `take 20 integers2` to evaluate the first 20 integers according to this definition. Describe the set which is actually defined by Attempt 2.

6.3.3   Third Attempt

The previous attempt gave a correct inductive definition of the integers, but there is still a problem with it, as can be seen by a simple example. Consider proving that -2 is in I:

1. $0 \in I$ base case
2. $0 \in I \rightarrow (1 \in I \land -1 \in I)$ instantiation, induction case
3. $1 \in I \land -1 \in I$ 1,2, *Modus Ponens*
4. $-1 \in I \rightarrow (0 \in I \land -2 \in I)$ instantiation, induction case
5. $0 \in I \land -2 \in I$ 3,4, *Modus Ponens*

We can be sure that this definition is correct because the induction step brings both the positive and negative numbers into I. Each element of I is incremented, guaranteeing that the naturals will be included as 1 and its successors are added. Each element of I is also decremented, ensuring that all of the negative numbers will be included as -1 and its predecessors are added.

A drawback, however, is that there are *two* ways for 0 to become a member of I. The base case puts it there, and step 5 does it again. In fact, the real defects of this definition are graphically illustrated by the fact that

```
take 20 integers2
 = [0,1,-1,2,0,0,-2,3,1,1,-1,1,-1,-1,-3,4,2,2,0,2]
```

It is more elegant if the definition introduces each element only once. Furthermore, practical software based in an inductively defined set would be inefficient (and possibly even incorrect) if the elements were introduced more than once. Here is one way to try to fix the problem:

Attempt 3. The set I is defined as follows:

- Base case: $0 \in I$

- Induction case: $x \in I \rightarrow (x + 1 \in I \land -(x + 1) \in I)$

- Extremal clause: nothing is in I unless its presence is justified by a finite number of uses of the base and induction cases.

Here is an implementation of the third definition:

```
nextIntegers3 :: Int -> [Int]
nextIntegers3 x = [x + 1, -(x + 1)]

integers3 :: [Int]
integers3 = builds 0 nextIntegers3
```

Exercise 15. Use the computer to examine the first 10 integers generated by this definition, and describe the set that is defined.

6.3.4   Fourth Attempt

This third attempt is a correct definition of the set of integers, and it is much closer to our intuition. What we want to do is say that if x is in I, then $-x$ is also in I. However, we also have to increment and decrement x somehow, so that all the integers can be included. Consider what happens when we attempt to show that -2 is in I:

1.	$0 \in I$	base case
2.	$0 \in I \rightarrow (1 \in I \wedge -1 \in I)$	instantiation, induction case
3.	$1 \in I \wedge -1 \in I$	1,2, *Modus Ponens*
4.	$1 \in I \rightarrow (2 \in I \wedge -2 \in I)$	instantiation, induction case
5.	$2 \in I \wedge -2 \in I$	3,4, and, *Modus Ponens*

This is almost what we want, except that when the induction rule is instantiated with -1 it places 0 in I again. And,

```
take 10 integers3 =
  [0,1,-1,2,-2,0,0,3,-3,-1].
```

Therefore Attempt 3 doesn't introduce each element precisely once.

Attempt 4. The set I of integers is defined as follows:

- Base case: $0 \in I$

- Induction case:

  1. $(x \in I \wedge x \geq 0) \rightarrow x + 1 \in I$

  2. $(x \in I \wedge x < 0) \rightarrow x - 1 \in I$

- Extremal clause: nothing is in I unless its presence is justified by a finite number of uses of the base and induction cases.

Here is an implementation of the fourth definition:

```
nextInteger4 :: Int -> Int
nextInteger4 x = if x < 0 then x - 1 else x + 1

integers4 :: [Int]
integers4 = build 0 nextInteger4
```

Exercise 16. Use the computer to generate some elements of the set defined by Attempt 4, and describe the result.

6.3.5 Fifth Attempt

Attempt 4 does not work, because the numbers generated from 0 must always be positive. It is necessary to go back to Attempt 4 and improve that. Furthermore, it would be better to have only one inductive case, if possible.

Attempt 5. The set I of integers is defined as follows:

- Base case: $0 \in I$

- Induction case: $(x \in I \land x \geq 0) \to x + 1 \in I \land -(x+1) \in I$

- Extremal clause: nothing is in I unless its presence is justified by a finite number of uses of the base and induction cases.

This definition is exactly what the original method of counting the integers suggested. It is implemented by the following Haskell program:

```
nextIntegers5 :: Int -> [Int]
nextIntegers5 x
   = if x > 0 \/ x == 0
         then [x + 1, -(x + 1)]
         else []

integers5 :: [Int]
integers5 = builds 0 nextIntegers5
```

Exercise 17. Use the computer to evaluate the first 10 elements of the set, and describe the result.

6.4 Suggestions for Further Reading

Many of the references on set theory cited in Chapter 4 also deal with inductive definitions of sets. *Elements of Set Theory*, by Enderton [10], gives some good examples of inductively defined sets. A more advanced treatment appears in *Axiomatic Set Theory*, by Suppes [28].

6.5 Review Exercises

Exercise 18. Does ints, using the following definition, enumerate the integers? If it does, then you should be able to pick any integer and see it eventually in the output produced by ints. Will you ever see the value -1?

```
nats :: [Int]
nats = build 0 (1 +)
```

```
negs :: [Int]
negs = build (-1) (1 -)

ints :: [Int]
ints = nats ++ negs
```

Exercise 19. Does twos enumerate the set of even natural numbers?

```
twos :: [Int]
twos = build 0 (2 *)
```

Exercise 20. What is wrong with the following definition of the stream of natural numbers?

```
nats = map (+ 1) nats ++ [0]
```

Exercise 21. What is the problem with the following definition of the naturals?

```
naturals :: [Int] -> [Int]
naturals (i:acc) = naturals (i + 1:i:acc)

nats :: [Int]
nats = naturals [0]
```

Exercise 22. Using induction, for any set S, define the powerset of S.

Exercise 23. Can we write a function that will take a stream of the naturals (appearing in any order) and give the index of a particular number?

Exercise 24. Using induction, define the set of roots of a given number n.

Exercise 25. Given the following definition, prove that n^3 is in set P of powers of n.

Definition 21. Given a number n, the set P of powers of n is defined as follows:

- $n^0 \in P$
- $n^m \in P \rightarrow n^{m+1} \in P$
- Nothing else is in P unless it can be shown to be in P by a finite number of uses of the base and induction rules.

Exercise 26. When is 0 in the set defined below?

Definition 22. Given a number n, the set N is defined as follows:

- $n \in N$
- $m \in N \rightarrow m - 2 \in N$

- Nothing is in N unless it can be shown to be in N by a finite number of uses of the previous rules.

Exercise 27. What set is defined by the following definition?

Definition 23. The set S is defined as follows:

- $1 \in S$
- $n \in S \land n \bmod 2 = 0 \rightarrow n + 1 \in S$
- $n \in S \land n \bmod 2 = 1 \rightarrow n + 2 \in S$
- Nothing else is in S unless it can be shown to be in S by a finite number of uses of the previous rules.

Exercise 28. Prove that 4 is in the set defined as follows:

Definition 24. The set S is defined as follows:

1. $0 \in S$
2. $n \in S \land n \bmod 2 = 0 \rightarrow n + 2 \in S$
3. $n \in S \land n \bmod 2 = 1 \rightarrow n + 1 \in S$
4. Nothing is in S unless it can be shown to be in S by a finite number of uses of the previous rules.

Exercise 29. Given the following definition, prove that the string 'yyyy' is in YYS. The symbol $\|$ here may mean concatenate or cons.

Definition 25. The set YYS of strings containing pairs of the letter 'y' is defined as follows:

1. "" $\in YYS$
2. $s \in YYS \rightarrow$ "yy" $\| s \in YYS$
3. Nothing else is in YYS unless it can be shown to be in YYS by a finite number of uses of rules (1) and (2).

Exercise 30. Using data recursion, define the set of strings containing the letter 'z'.

Exercise 31. Using induction, define the set of strings of spaces of length less than or equal to some positive integer n.

Exercise 32. Using recursion, define the set of strings of spaces of length less than or equal to length n, where n is a positive integer.

Exercise 33. Using induction, define the set of sets of integers SSI, each of which is missing a distinct natural number.

Exercise 34. Given the following definition, show that the set $I - \{-3\} \in SSI-$.

The set of sets of integers SSI, each of which is missing a distinct negative integer, is defined inductively as follows:

1. $I - \{-1\} \in SSI-$

2. $I - \{n\} \rightarrow I - \{n-1\} \in SSI-$

3. Nothing else is in SSI- unless it can be shown to be in SSI- by a finite number of uses of rules (1) and (2).

Exercise 35. Given the following definition, prove that -7 is in ONI. The set ONI of odd negative integers is defined as follows:

1. $-1 \in ONI$

2. $n \in ONI \rightarrow n - 2 \in ONI$

3. Nothing is in ONI unless it can be shown to be in ONI by a finite number of uses of the previous rules.

Exercise 36. Using data recursion in Haskell, define the set `ni` of negative integers.

Exercise 37. If you print the elements of

```
[(a,b) | a <- [0..], b <- [0..]]
```

will you ever see the element (1,2)?

Exercise 38. What set is given by the following definition?

Definition 26. The set S is defined as follows:

1. $1 \in S$

2. $n \in S \rightarrow n - n \in S$

3. Nothing is in S unless it can be shown to be in S by a finite number of uses of the previous rules.

Chapter 7

Induction

A common type of problem is to prove that an object x has some property P. The mathematical notation for this is $P(x)$, where P stands for predicate (or property). For example, if x is 6 and $P(x)$ is the predicate 'x is an even number', then we could express the statement '6 is an even number' with the shorthand mathematical statement $P(6)$.

Many computer science applications require us to prove that *all* the elements of a set S have a certain property P. This can be written as

$$\forall x \in S \,.\, P(x).$$

Statements like this can be used to assert properties of data: for example, we could state that every item in a database has a certain property. They can also be used to describe the behaviour of computer programs. For example, $P(n)$ might be a statement that a loop computes the correct result if it terminates after n iterations, and the statement $\forall n \in N \,.\, P(n)$ says that the loop is correct if it terminates after *any* number of iterations.

One approach to proving an assertion about all the elements of a set is to write out a separate direct proof for each element of the set. That would be all right if the set were small. For example, to prove that all the elements of the set $\{4, 6\}$ are even, you could just prove that 4 is even and also that 6 is even. However, this direct approach quickly becomes tedious for large sets: to prove that all the elements of $\{2, 4, 6, \ldots, 1000\}$ are even, you would need 500 separate proofs! Even worse, the brute-force method doesn't work at all—even in principle—for infinite sets, because proofs are always required to be finitely long.

Induction is a powerful method for proving that every element of a set has a certain property, and it is used frequently in computer science applications. This chapter shows you how inductive proofs work, and gives many examples, including both mathematical and computing applications.

7.1   The Principle of Mathematical Induction

Induction is used to prove that a property $P(x)$ holds for every element of a set S. Instead of proving $P(x)$ separately for every x, induction relies on a systematic procedure: you just have to prove a few facts about the set S and the property P, and then a theorem called the Principle of Mathematical Induction allows you to conclude $\forall x \in S \,.\, P(x)$.

The basic idea is to define a systematic procedure for proving that P holds for an arbitrary element. To do this, all the elements of the set must first be organised into a *chain*. A chain is a sequence of elements with two properties: there is a starting point called the *base element*, and each element of the chain has exactly one successor element. In effect, this simply means that it is possible to enumerate all the elements of S as a sequence of indexed elements

$$S = \{x_0, x_1, x_2, \dots\},$$

where x_0 is the base element and the successor of x_i is x_{i+1}. This is not a trivial restriction: for instance, it is impossible to enumerate all the real numbers in this way, so mathematical induction cannot be used to prove properties of the real numbers. However, the natural numbers and most of the objects that we use in computing applications can easily be organised into a chain.

Once the set is organised as a chain, we must prove two statements:

1. The *base case* $P(x_0)$ says that the property P holds for the base element x_0.

2. The *inductive case* $P(x_i) \to P(x_{i+1})$ says that if P holds for an arbitrary element x_i of the set, then it must also hold for the successor element x_{i+1}.

Since every element of the set is in the chain, this means you can establish that P holds for a particular element using a finite sequence of logical inferences. For example, $P(x_4)$ can be proved using the following steps:

Conclusion	Justification		
$P(x_0)$	the base case		
$P(x_1)$	$P(x_0) \to P(x_1)$	$\wedge$	$P(x_0)$
$P(x_2)$	$P(x_1) \to P(x_2)$	$\wedge$	$P(x_1)$
$P(x_3)$	$P(x_2) \to P(x_3)$	$\wedge$	$P(x_2)$
$P(x_4)$	$P(x_3) \to P(x_4)$	$\wedge$	$P(x_3)$

Given any element x_k of S, you can prove $P(x_k)$ using this strategy, and the proof will be $k + 1$ lines long. We have not yet used the Principle of Mathematical Induction; we have just used ordinary propositional calculus. However, proving $P(x_k)$ for an arbitrary k is not the same as proving $\forall k \in N \,.\, P(x_k)$, because there may be an infinite number of values of k, so the proof would be infinitely long. The size of a proof must always be finite.

What the Principle of Mathematical Induction allows us to conclude is that P holds for all the elements of S, even if there is an infinite number of them. Thus it introduces something new—it isn't just a macro that expands out into a long proof.

Theorem 39 (Principle of Mathematical Induction). Let the set S be a chain $x_0, x_1, \ldots$ with base element x_0, and let $P(x)$ be a predicate which is either true or false for every element of S. If $P(x_0)$ is true and $P(x_n) \to P(x_{n+1})$ holds for all $n \geq 0$, then $\forall x \in S \ . \ P(x)$ is true.

This theorem can be proved using axiomatic set theory, which is beyond the scope of this book. For most applications in computer science it is sufficient to have an intuitive understanding of what the theorem says, and to have a working understanding of how to use it in proving other theorems.

There is a strong similarity between an inductive definition of a set and an inductive proof. Each element of a set defined by induction is shown to be in the set by a finite number of applications of the induction and base steps; inductive proofs have a similar structure, with a base case and an induction step.

7.2 Induction on Natural Numbers

There is a traditional story about Gauss, one of the greatest mathematicians in history. In school one day the teacher told the class to work out the sum $1 + 2 + \cdots + 100$. After a short time thinking, Gauss gave the correct answer— 5050—long before it would have been possible to work out all the additions. He had noticed that the sum can be arranged into pairs of numbers, like this:

$$(1 + 100) + (2 + 99) + (3 + 98) + \cdots + (50 + 51)$$

The total of each pair is 101, and there are 50 of the pairs, so the result is $50 \times 101 = 5050$.

Methods like this can often be used to save time in computing, so it is worthwhile to find a solution to the general case of this problem, which is the sum

$$\sum_{i=1}^{n} i = 1 + 2 + \cdots + n.$$

If n is even, then we get $\frac{n}{2}$ pairs that all total to $n+1$, so the result is $\frac{n}{2} \times (n+1) = \frac{n \times (n+1)}{2}$. If n is odd, we can start from 0 instead of 1; for example,

$$\sum_{i=0}^{7} i = (0 + 7) + (1 + 6) + (2 + 5) + (3 + 4).$$

Figure 7.1: Geometric Interpretation of Sum

In this case there are $\frac{n+1}{2}$ pairs each of which totals to n, so the result is again $\frac{n \times (n+1)}{2}$. For any natural number n, we end up with the result

$$\sum_{i=0}^{n} i = \frac{n \times (n+1)}{2}.$$

This formula is useful because it reduces the computation time required. For example, if n is 1000 then it would take 1000 additions to work out the summation by brute force, but the formula always requires just one addition, one multiplication and one division.

Figure 7.1 shows another way to understand the formula. The rectangle is covered half by dots and half by stars. The number of stars is $1+2+3+4$, and the area of the rectangle is $4 \times (4+1)$, so the total number of stars is $\frac{4 \times (4+1)}{2}$.

So far we have only guessed the general formula for $\sum_{i=0}^{n} i$, and we have considered two ways of understanding it. The next step is to *prove* that this formula always gives the right answer, and induction provides the way to do it.

Theorem 40.

$$\forall n \in N \, . \, \sum_{i=0}^{n} i = \frac{n \times (n+1)}{2}$$

Proof. Define the property P as follows:

$$P(n) = \left(\sum_{i=0}^{n} i = \frac{n \times (n+1)}{2} \right)$$

Thus the aim is to prove that

$$\forall n \in N \, . \, P(n)$$

and we proceed by induction on n.

Base case. We need to prove $P(0)$.

$$\sum_{i=0}^{0} i = 0$$
$$= \frac{0 \times (0+1)}{2}$$

Thus we have established the property $P(0)$.

Induction case. Assume for a particular n that

$$\sum_{i=0}^{n} i = \frac{n \times (n+1)}{2}.$$

The aim is to show (for this particular value of n) that

$$\sum_{i=0}^{n+1} i = \frac{(n+1) \times (n+2)}{2}.$$

We do this by starting with the left hand side of the equation and using algebra to transform it into the right hand side. The first step uses the assumption given above. This is called the *induction hypothesis.*

$$
\begin{aligned}
\sum_{i=0}^{n+1} i &= \sum_{i=0}^{n} i + (n+1) \\
&= \frac{n \times (n+1)}{2} + (n+1) \\
&= \frac{n \times (n+1)}{2} + \frac{2 \times (n+1)}{2} \\
&= \frac{n \times (n+1) + 2 \times (n+1)}{2} \\
&= \frac{(n+1) \times (n+2)}{2}
\end{aligned}
$$

Use the principle of induction. Now we have established that

$$\left(\sum_{i=0}^{n} i = \frac{n \times (n+1)}{2} \right) \rightarrow \left(\sum_{i=0}^{n+1} i = \frac{(n+1) \times (n+2)}{2} \right).$$

which means that

$$P(n) \rightarrow P(n+1).$$

According to the principle of induction, the base case, $P(0)$, and this logical implication together imply that the theorem is true. □

Exercise 1. Let a be an arbitrary real number. Prove, for all natural numbers m and n, that $a^{m \times n} = (a^m)^n$.

Exercise 2. Prove that the sum of the first n odd positive numbers is n^2.

7.3   Induction and Recursion

Induction is a common method for proving properties of recursively defined functions. The *factorial* function, which is defined recursively, provides a good example of an induction proof.

> *factorial* :: *Natural* → *Natural*
> *factorial* 0 = 1
> *factorial* (n + 1) = (n + 1) * *factorial* n

A common problem in computer science is to prove that a program computes the correct answer. Such a theorem requires an abstract mathematical specification of the problem, and it has to show that for all inputs, the program produces the same result that is defined by the specification. The value of n! (factorial of n) is defined as the product of all the natural numbers from 1 through n; the standard notation for this is $\prod_{i=1}^{n} i$. The following theorem says that the *factorial* function as defined above actually computes the value of n!.

Theorem 41. For all natural numbers n, *factorial* n $= \prod_{i=1}^{n} i$.

Proof. Induction on n.

> *factorial* 0
> $= 1$ *factorial*.1
> $= \prod_{1}^{0} i$ def. of $\prod$

For the induction case, it is necessary to prove that

$$\left(factorial\ n\ =\ \prod_{i=1}^{n} i \right) \rightarrow \left(factorial\ (n+1)\ =\ \prod_{i=1}^{n+1} i \right).$$

This is done by assuming the left side as the inductive hypothesis: for a particular n, the hypothesis is *factorial* n $= \prod_{i=1}^{n} i$. Given this assumption, we must prove that *factorial* (n + 1) $= \prod_{i=1}^{n+1} i$.

> *factorial* (n + 1)
> $= (n + 1) \times factorial\ n$ *factorial*.2
> $= (n + 1) \times \prod_{i=1}^{n} i$ hypothesis
> $= \prod_{i=1}^{n+1} i$ def. $\prod$

$\square$

7.4   Induction on Peano Naturals

The Peano representation of natural numbers is a rich source of examples for induction. Actually, it's hard to do anything at all in Peano arithmetic without induction. For example, how do we even know that a natural number is equal to itself? The following theorem says so, and its proof requires induction.

Theorem 42 (Self equality). $\forall x :: Nat.\ equals\ x\ x = True$

Proof. Base case:

$$equals\ Zero\ Zero$$
$$=\ True \qquad\qquad\qquad\qquad equals.1$$

For the inductive case, assume that $equals\ x\ x = True$. We must prove that $equals\ (Succ\ x)\ (Succ\ x) = True$.

$$equals\ (Succ\ x)\ (Succ\ x)$$
$$=\ equals\ x\ x \qquad\qquad\qquad\qquad equals.2$$
$$=\ True \qquad\qquad\qquad\qquad\qquad \text{hypothesis}$$

□

This proof illustrates a subtle issue. When we are using the languages of English and mathematics to talk *about* natural numbers, we can assume that anything is equal to itself. We don't normally prove theorems like $x = x$. However, that is not what we have just proved: the theorem above says that x is the same as itself according to *equals*, which is defined *inside* the Peano system. Therefore the proof also needs to work inside the Peano system; hence the induction.

We'll look at a few more typical Peano arithmetic theorems, both to see how the Peano natural numbers work and to get more practice with induction. The following theorem says, in effect, that $(x + y) - x = y$. In elementary algebra, we would prove this by calculating $(x + y) - x = (x - x) + y = 0 + y = y$. The point here, however, is that the addition and subtraction are being performed by the recursive *add* and *sub* functions, and we can use those to prove the theorem directly.

Theorem 43. $sub\ (add\ x\ y)\ x = y$

Proof. Induction over x. The base case is

$$sub\ (add\ Zero\ y)\ Zero$$
$$=\ sub\ y\ Zero \qquad\qquad\qquad\qquad add.1$$
$$=\ y \qquad\qquad\qquad\qquad\qquad\quad sub.1$$

For the inductive case, assume $sub\ (add\ x\ y)\ x = y$; the aim is to prove $sub\ (add\ (Succ\ x)\ y)\ (Succ\ x) = y$.

$sub\ (add\ (Succ\ x)\ y)\ (Succ\ x)$
  $= sub\ (Succ\ (add\ x\ y))\ (Succ\ x)$ $add.2$
  $= sub\ (add\ x\ y)\ x$ $sub.3$
  $= y$ hypothesis

□

The proof above happens to go through directly and easily, but many simple theorems do not. For example, it is considerably harder to prove $(x+y)-y = x$ than to prove $(x + y) - x = y$. Even worse, there is no end to such theorems. Instead of continuing to choose theorems based on their ease of proof, it is better to proceed systematically by developing the standard properties of natural numbers, such as associativity and commutativity. The attractive feature of the Peano definitions is that all these laws are theorems; the only definitions we need are the basic ones given already. It is straightforward to prove that addition is associative, so we begin with that property.

Theorem 44 (add is associative). $add\ x\ (add\ y\ z)\ =\ add\ (add\ x\ y)\ z$

Proof. Induction over x. The Base case is

$add\ Zero\ (add\ y\ z)$
  $= add\ y\ z$ $add.1$
  $= add\ (add\ Zero\ y)\ z$ $add.1$

Inductive case. Assume $add\ x\ (add\ y\ z)\ =\ add\ (add\ x\ y)\ z$. Then

$add\ (Succ\ x)\ (add\ y\ z)$
  $= Succ\ (add\ x\ (add\ y\ z))$ $add.2$
  $= Succ\ (add\ (add\ x\ y)\ z)$ hypothesis
  $= add\ (Succ\ (add\ x\ y))\ z$ $add.2$
  $= add\ (add\ (Succ\ x)\ y)\ z$ $add.2$

□

Next, it would be good to prove that addition is commutative: $x + y = y + x$. To prove this, however, a sequence of simpler theorems is needed—each providing yet another example of induction.

First we need to be able to simplify additions where the *second* argument is Zero. We know already that $add\ Zero\ x\ =\ x$; in fact, this is one of the Peano axioms. It is *not* an axiom that $add\ x\ Zero\ =\ x$; that is a theorem requiring proof.

Theorem 45. $add\ x\ Zero\ =\ x$

Proof. Induction over x. The base case is

$add\ Zero\ Zero$
  $= Zero$ $add.1$

For the inductive case, we assume $add\ x\ Zero\ =\ x$. Then

$add\ (Succ\ x)\ Zero$
$\quad =\ Succ\ (add\ x\ Zero)$ add.2
$\quad =\ Succ\ x$ hypothesis

                     □

    The next theorem allows us to move a `Succ` from one argument of an addition to the other. It says, in effect, that $(x + 1) + y = x + (y + 1)$. That may not sound very dramatic, but many proofs require the ability to take a little off one argument and add it onto the other.

Theorem 46. $add\ (Succ\ x)\ y\ =\ add\ x\ (Succ\ y)$

Proof. Induction over x. Base case:

$add\ (Succ\ Zero)\ y$
$\quad =\ Succ\ (add\ Zero\ y)$ add.2
$\quad =\ Succ\ y$ add.1
$\quad =\ add\ Zero\ (Succ\ y)$ add.1

Inductive case:

$add\ (Succ\ (Succ\ x))\ y$
$\quad =\ Succ\ (add\ (Succ\ x)\ y)$ add.2
$\quad =\ Succ\ (add\ x\ (Succ\ y))$ hypothesis
$\quad =\ add\ (Succ\ x)\ (Succ\ y)$ add.2

                     □

    Now we can prove that Peano addition is commutative.

Theorem 47 (add is commutative). $add\ x\ y\ =\ add\ y\ x$

Proof. Induction over x. Base case:

$add\ Zero\ y$
$\quad =\ y$ add.1
$\quad =\ add\ y\ Zero$ Theorem 45

Inductive case: assume that $add\ x\ y\ =\ add\ y\ x$. Then

$add\ (Succ\ x)\ y$
$\quad =\ Succ\ (add\ x\ y)$ add.2
$\quad =\ Succ\ (add\ y\ x)$ hypothesis
$\quad =\ add\ (Succ\ y)\ x$ add.2
$\quad =\ add\ y\ (Succ\ x)$ Theorem 46

                     □

7.5 Induction on Lists

Lists are one of the most commonly used data structures in computing, and there is a large family of functions to manipulate them. These functions are typically defined recursively, with a base case for empty lists [] and a recursive case for non-empty lists of the form $(x : xs)$. Induction is the most common method for proving properties of such functions.

Before going on, we discuss some practical techniques that help in coping with theorems. If you aren't familiar with them, mathematical statements sometimes look confusing, and it is easy to develop a bad habit of skipping over the mathematics when reading a book. Here is some advice on better approaches; we will try to illustrate the advice in a concrete way in this section, but these methods will pay off throughout your work in computer science, not just in the section you're reading right now.

- When you are faced with a new theorem, try to understand what it means before trying to prove it. Restate the main idea in English.

- Think about what applications the theorem might have. If it says that two expressions are the same, can you think of situations where there might be a practical advantage in replacing the left hand side by the right hand side, or vice versa? (A common situation is that one side of the equation is more natural to write, and the other side is more efficient.)

- Try out the theorem on some small examples. Theorems are often stated as equations; make up some suitable input data and evaluate both sides of the equation, to see that they are the same.

- Check what happens in boundary cases. If the equation says something about lists, what happens if the list is empty? What happens if the list is infinite?

- Does the theorem seem related to other ones? Small theorems about functions—the kind that are usually proved by induction—tend to fit together in families. Noticing these relationships helps in understanding, remembering and applying the results.

Now, bearing these ideas in mind, we will work through a series of examples where induction is used to prove theorems about the properties of recursive functions over lists.

Induction over lists is used to prove that a proposition $P(xs)$ holds for every list xs (such that $P(xs)$ has a valid type). The base case is to prove that P holds for the empty list: $P([])$. The inductive case is to prove that if P holds for a list xs, then it must also hold for any list of the form $x : xs$. When the base and inductive case are established, then the principle of induction allows us to conclude that $P(xs)$ holds for every list xs which has finite length. (We discuss infinite lists in Section 7.9).

The *sum* function takes a list of numbers and adds them up. It defines the sum of an empty list to be 0, and the sum of a non-empty list is computed by adding the first number onto the sum of all the rest.

```
sum :: Num a => [a] -> a
sum [] = 0
sum (x:xs) = x + sum xs
```

The following theorem states a useful fact about the relationship between two functions: *sum* and ++. It says that if you have two lists, say *xs* and *ys*, then there are two different ways to compute the combined total of all the elements. You can either append the lists together with ++, and then apply *sum*, or you can apply *sum* independently to the two lists, and add the two resulting numbers.

Theorems of this sort are often useful for transforming programs to make them more efficient. For example, suppose that you have a very long list to sum up, and two computers are available. We could use parallelism to cut the execution time almost in half. The idea is to split up the long list into two shorter ones, which can be summed in parallel. One quick addition will then suffice to get the final result. This technique is an example of the *divide and conquer* strategy for improving the efficiency of algorithms. Obviously there is more to parallel computing and program optimisation than we have covered in this paragraph, but theorems like the one we are considering really do have practical applications.

Theorem 48. $sum\ (xs + ys) = sum\ xs + sum\ ys$

The proof of this theorem is a typical induction over lists, and it provides a good model to follow for future problems.

Proof. Induction over *xs*. The base case is

$$sum\ ([]\ + ys)$$
$$= sum\ ys \qquad\qquad\qquad (+).1$$
$$= 0 + sum\ ys \qquad\qquad\quad 0 + x = x$$
$$= sum\ []\ + sum\ ys \qquad\quad sum.1$$

The inductive case is

$$sum\ ((x:xs) + ys)$$
$$= sum\ (x:(xs + ys)) \qquad\qquad append.2$$
$$= x + sum\ (xs + ys) \qquad\qquad sum.2$$
$$= x + (sum\ xs + sum\ ys) \qquad\quad \text{hypothesis}$$
$$= (x + sum\ xs) + sum\ ys \qquad\quad +. \text{ is associative}$$
$$= sum\ (x:xs) + sum\ ys \qquad\qquad sum.2$$

$\square$

Many theorems describe a relationship between two functions; the previous one is about the combination of *sum* and (++), while this one combines *length* with (++). Its proof is left as an exercise.

Theorem 49. $length\ (xs{+}{+}ys)\ =\ length\ xs\ +\ length\ ys$

We now consider several theorems that state crucial properties of the *map* function. These theorems are important in their own right, and they are commonly used in program transformation and compiler implementation. They are also frequently used to justify steps in proofs of more complex theorems.

The following theorem says that *map* is 'length preserving': when you apply *map* to a list, the result has the same length as the input list.

Theorem 50. $length\ (map\ f\ xs) = length\ xs$

Proof. Induction over xs. The base case is

$length\ (map\ f\ [])$
$\quad = length\ []$ $map.1$

For the inductive case, assume $length\ (map\ f\ xs) = length\ xs$. Then

$length\ (map\ f\ (x : xs))$
$\quad = length\ (f\ x\ :\ map\ f\ xs)$ $map.2$
$\quad = 1 + length\ (map\ f\ xs)$ $length.2$
$\quad = 1 + length\ xs$ hypothesis
$\quad = length\ (x : xs)$ $length.2$

$\square$

The next theorem is reminiscent of Theorem 48; it says you can get the same result by either of two methods: (1) mapping a function over two lists and then appending the results together, and (2) appending the input lists and then performing one longer map over the result. Its proof is yet another good example of induction, and is left as an exercise.

Theorem 51. $map\ f\ (xs{+}{+}ys)\ =\ map\ f\ xs\ {+}{+}\ map\ f\ ys$

  One of the most important properties of map is expressed precisely by the following theorem. Suppose that you have two computations to perform on all the elements of a list. First you want to apply g to an element, getting an intermediate result to which you want to apply f. There are two methods for doing the computation. The first method uses two separate loops, one to perform g on every element and the second loop to perform f on the list of intermediate results. The second method is to use just one loop, and each iteration performs the g and f applications in sequence. This theorem is used commonly by optimising compilers, program transformations (both manual and automatic), and it's also vitally important in parallel programming. Again, we leave the proof as an exercise.

Theorem 52. $(map\ f\ .\ map\ g)\ xs\ =\ map\ (f.g)\ xs$

For a change of pace, we now consider an intriguing theorem. Once you understand what it says, however, it becomes perfectly intuitive.

Theorem 53. $sum\ (map\ (1+)\ xs)\ =\ length\ xs\ +\ sum\ xs$

Proof. Induction over xs. The base case is

$sum\ (map\ (1+)\ [])$
$= sum\ []$ $map.1$
$= 0 + sum\ []$ $0 + x = x$
$= length\ []\ +\ sum\ []$ $length.1$

For the inductive case, assume $sum\ (map\ (1+)\ xs)\ =\ length\ xs\ +\ sum\ xs$. Then

$sum\ (map\ (1+)\ (x:xs))$
$= sum\ ((1+x)\ :\ map\ (1+)\ xs)$ $map.2$
$= (1+x)\ +\ sum\ (map\ (1+)\ xs)$ $sum.2$
$= (1+x)\ +\ (length\ xs\ +\ sum\ xs)$ hypothesis
$= (1+length\ xs)\ +\ (x\ +\ sum\ xs)$ $(+).algebra$
$= length\ (x:xs)\ +\ sum\ (x:xs)$ $length.2,\ sum.2$

□

The *foldr* function is important because it expresses a basic looping pattern. There are many important properties of *foldr* and related functions. Here is one of them:

Theorem 54. $foldr\ (:)\ []\ xs\ =\ xs$

Some of the earlier theorems may be easy to understand at a glance, but that is unlikely to be true for this one! Recall that the *foldr* function takes apart a list, and combines their elements using an operator. For example,

$$foldr\ (+)\ 0\ [1,2,3]\ =\ 1\ +\ (2\ +\ (3\ +\ 0))$$

Now, what happens if we combine the elements of the list using the cons operator $(:)$ instead of addition, and if we use $[]$ as the initial value for the recursion instead of 0? The previous equation then becomes

$$foldr\ (:)\ []\ [1,2,3]\ =\ 1\ :\ (2\ :\ (3\ :\ []))$$
$$=\ [1,2,3].$$

We ended up with the same list we started out with, and the theorem says this will always happen, not just with the example $[1,2,3]$ used here.

Proof. Induction over xs. The base case is

$foldr\ (:)\ []\ []$
    $= []$ $foldr.1$

Now assume that $foldr\ (:)\ []\ xs\ =\ xs$; then the inductive case is

$foldr\ (:)\ []\ (x:xs)$
    $= x\ :\ foldr\ (:)\ []\ xs$ $foldr.2$
    $= x:xs$ hypothesis

☐

Suppose you have a list of lists, of the form $xss\ =\ [xs_0, xs_1, \ldots, xs_n]$. All of the lists xs_i must have the same type $[a]$, and the type of xss is $[[a]]$. We might want to apply a function $f :: a \to b$ to all the elements of all the lists, and build up a list of all the results. There are two different ways to organise this computation:

- Use the *concat* function to make a single flat list of type $[a]$ containing all the values, and then apply *map* f to produce the result with type $[b]$.

- Apply *map* f separately to each xs_i, by computing *map* (*map* f) xss, producing a list of type $[[b]]$. Then use *concat* to flatten them into a single list of type $[b]$.

The following theorem guarantees that both approaches produce the same result. This is significant because there are many practical situations where it is more convenient to write an algorithm using one approach, yet the other is more efficient.

Theorem 55. $map\ f\ (concat\ xss)\ =\ concat\ (map\ (map\ f)\ xss)$

Proof. Proof by induction over xss. The base case is

$map\ f\ (concat\ [])$
    $=\ map\ f\ []$ $concat.1$
    $=\ []$ $map.1$
    $=\ concat\ []$ $concat.1$
    $=\ concat\ (map\ (map\ f)\ [])$ $map.1$

Assume that $map\ f\ (concat\ xss)\ =\ concat\ (map\ (map\ f)\ xss)$. The inductive case is

$map\ f\ (concat\ (xs:xss))$
    $=\ map\ f\ (xs \mathbin{+\!\!+} concat\ xss)$ $concat.2$
    $=\ map\ f\ xs \mathbin{+\!\!+} map\ f\ (concat\ xss)$ Theorem 51
    $=\ map\ f\ xs \mathbin{+\!\!+} concat\ (map\ (map\ f)\ xss)$ hypothesis
    $=\ concat\ (map\ f\ xs\ :\ map\ (map\ f)\ xss)$ $concat.2$
    $=\ concat\ (map\ (map\ f)\ (xs:xss))$ $map.2$

☐

Sometimes you don't need to perform an induction, because a simpler proof technique is already available. Here is a typical example:

Theorem 56. *length* $(xs{+}{+}(y:ys)) = 1 + length\ xs + length\ ys$

This theorem could certainly be proved by induction (and that might be good practice for you!) but we already have a similar theorem which says that *length* $(xs{+}{+}ys) = length\ xs + length\ ys$. Instead of starting a new induction completely afresh, it's more elegant to carry out a few steps that enable us to apply the existing theorem. Just as reuse of software is a good idea, reuse of theorems is good style in theoretical computer science.

$$length\ (xs + (y:ys))$$
$$= length\ xs + length\ (y:ys)$$
$$= length\ xs + (1 + length\ ys)$$
$$= 1 + length\ xs + length\ ys$$

Exercise 3. Prove Theorem 49.

Exercise 4. Prove Theorem 51.

Exercise 5. Prove Theorem 52.

7.6 Functional Equality

Many theorems used in computer science (including most of the ones in this chapter) say that two different algorithms are always guaranteed to produce the same result. The algorithms are defined as functions, and the theorem says that when you apply two different functions to the same argument, they give the same result.

It is simpler and more direct simply to say that the two functions are equal. However, this raises an interesting question: what does it mean to say $f = g$ when f and g are functions? (This issue will be revisited in Chapter 9.) There are at least two standard notions of functional equality that are completely different from each other, so it pays to be careful!

- *Intensional equality.* Two functions f and g are intensionally equal if their definitions are identical. This means, of course, that the functions are not equal if their types are different. If they are computer programs, testing for intensional equality involves comparing the source programs, character by character. The functions are intensionally equal if their definitions *look* the same.

- *Extensional equality.* Two functions f and g are extensionally equal if they have the same type $a \rightarrow b$ and $f\ x = g\ x$ for all well typed arguments $x :: a$. More precisely, $f = g$ if and only if

$$\forall x :: a \ . \ f\ x \ = \ g\ x.$$

The functions are extensionally equal if their definitions *behave* the same.

In computer science we are almost always interested in extensional equality. A typical situation is that we have an algorithm, expressed as a function f, and the aim is to replace it by a more efficient function g. This will not affect the correctness of the program as long as f and g are extensionally equal, but they are obviously not intensionally equal.

Some of the theorems given in the previous section can be stated in a simpler fashion using extensional equality. For example, recall Theorem 50, which says that map doesn't change the length of its argument:

$$length\ (map\ f\ xs) = length\ xs$$

A more direct way to state the same fact is to omit the irrelevant argument xs, and just say that these two functions are equal:

$$length\ .\ (map\ f) = length$$

To prove such a theorem of the form $f = g$, we need only prove that $\forall x :: a.\ f\ x = g\ x$, and this can be achieved by choosing an arbitrary $x :: a$, and proving the equation $f\ x = g\ x$.

Theorem 57. $foldr\ (:)\ [\ ] \ = \ id$

Proof. The equation states that two functions are equal: the right hand side, *id*, is a function, and the left hand side is a partial application (*foldr* takes three arguments, but it has been applied to only two), so that is also a function. Therefore we use the definition of extensional equality of functions; thus we choose an arbitrary list xs, and we must prove that $foldr\ (:)\ [\ ]\ xs \ = \ id\ xs \ = \ xs$. Now the right hand side is just xs, by the definition of *id*, so the equation is proved by Theorem 54.   □

Theorem 58. $map\ f\ .\ concat \ = \ concat\ (map\ (map\ f))$.

Exercise 6. Prove Theorem 58.

Exercise 7. Prove that the $+\!\!+$ operator is associative.

7.7 Induction on Trees

Induction can be generalised to work for tree structures. The method is similar to lists, except that the base case is used for leaves, and the inductive case is used for nodes. There is only one list data structure, so all the list functions are well standardised. However, there are many different ways to define trees. It is common for a new software project to require a slightly different kind of tree, and reasoning about the software will therefore require new theorems to be proved by induction.

To illustrate how to write induction proofs for theorems about trees, we will use a data type definition where the leaf nodes have no locally stored data—they are simply indicated by the *Tip* constant—and a *Node* contains an integer and two subtrees.

```
data Tree = Tip | Node Int Tree Tree
```

The *reflect* function takes a tree and returns its mirror image, where everything is reversed left-to-right.

```
reflect :: Tree -> Tree
reflect Tip = Tip
reflect (Node n l r) = Node n (reflect r) (reflect l)
```

The following theorem says that if you reflect a finite tree twice, you get the same tree back.

Theorem 59. *reflect (reflect t) = t*

Proof. Let the proposition $P(t) = (reflect\ (reflect\ t) = t)$. The theorem is proved by induction over t. The base case is $P(Tip)$:

$\quad$ *reflect (reflect Tip)*
$\quad = reflect\ Tip$
$\quad = \quad Tip$

Inductive case. Let $l, r :: Tree$ be trees, and assume $P(l)$ and $P(r)$. The aim is to prove $P(Node\ x\ l\ r)$ for arbitrary $x :: Int$.

$\quad$ *reflect (reflect (Node x l r))*
$\quad = reflect\ (Node\ x\ (reflect\ r)\ (reflect\ l))$
$\quad = Node\ x\ (reflect\ (reflect\ l))(reflect\ (reflect\ r))$
$\quad = Node\ x\ l\ r$

$\hfill\square$

Some of the most important properties of trees are concerned with the numbers of nodes in the two branches, and with the heights of trees and subtrees. The time required by many algorithms depends on the heights of trees, so the science of algorithmics is often concerned with these properties.

```
height :: Tree -> Int
height Tip = 0
height (Node x l r) = 1 + max (height l) (height r)

balanced :: Tree -> Bool
balanced Tip = True
balanced (Node x l r) =
   balanced l && balanced r && (height l == height r)
nodecount :: Tree -> Int
nodecount Tip = 0
nodecount (Node x l r) = 1 + nodecount l + nodecount r
```

The following theorem gives the number of nodes in a tree, provided that tree is balanced.

Theorem 60. Let $h = height\ t$. If $balanced\ t$, then $nodecount\ t = 2^h - 1$.

Proof. Let $P(t) = balanced\ t \rightarrow nodecount\ t = 2^h - 1$. We prove $P(t)$ by induction over the tree structure. For the base case, we need to prove that the theorem holds for a *Tip*.

$balanced\ Tip = True$
$height\ Tip = 0$
$2^h - 1 = 0$
$nodecount\ Tip = 0$

For the inductive case, let $t = Node\ x\ l\ r$, and let $hl = height\ l$ and $hr = height\ r$. Assume $P(l)$ and $P(r)$; the aim is to prove $P(t)$. There are two cases to consider. If t is not balanced, then the implication $balanced\ t \rightarrow P(t)$ is vacuously true. If t is balanced, however, then the implication is true if and only if $P(t)$ is true. Therefore we need to prove $P(t)$ given the following three assumptions: (1) $P(l)$ (inductive hypothesis), (2) $P(r)$ (inductive hypothesis), and $balanced\ t$ (premise of implication to be proved).

$$
\begin{aligned}
h &= height\ (Node\ x\ l\ r) \\
&= 1 + \max\ (height\ l)\ (height\ r) && height.2 \\
&= 1 + height\ l && \text{assumption} \\
&= 1 + hl && \text{def } hl \\
nodecount\ t & \\
&= nodecount\ (Node\ x\ l\ r) && \text{def } t \\
&= 1 + nodecount\ l\ +\ nodecount\ r && nodecount.2 \\
&= 1 + 2^{hl} - 1 + 2^{hr} - 1 && \text{hypothesis} \\
&= 2^{hl} + 2^{hr} - 1 && \text{arithmetic} \\
&= 2^{hl} + 2^{hl} - 1 && hl = hr \\
&= 2 \times 2^{hl} - 1 && \text{algebra} \\
&= 2^{hl+1} - 1 && \text{algebra} \\
&= 2^h - 1 && \text{def } h
\end{aligned}
$$

□

7.8   Pitfalls and Common Mistakes

After a bit of practice, induction can come to seem almost too easy. You just
set up the base and inductive cases, crank the handle, and out comes a proof.

There are many kinds of bad inductive proofs. Their flaws are often due to
suspicious base cases, although there are a variety of dubious ways in which to
prove the induction cases too. Here is an interesting example.

7.8.1   A Horse of Another Colour

The following theorem is a famous classic.

Theorem 61. All horses are the same colour.

Proof. Define $P(n)$ to mean 'in any set containing n horses, all of them have
the same colour'. We proceed by induction over n.

Base case. Every horse has the same colour as itself, so $P(1)$ is true.

Inductive case. Assume $P(n)$, and consider a set containing $n + 1$ horses;
call them $h_1, h_2, \ldots, h_{n+1}$. We can define two subsets $A = h_1, \ldots, h_n$ and
$B = h_2, \ldots h_{n+1}$. Both sets A and B contain n horses, so all the horses in A
are the same colour (call it C_A), and all the horses in B are the same colour (call
it C_B). Pick one of the horses that is an element of both A and B. Clearly this
horse has the same colour as itself; call it C_h. Thus $C_A = C_h = C_B$. Therefore
all the horses $h_1, \ldots, h_{n+1}$ have the same colour.

Since we have proved $P(n) \rightarrow P(n+1)$, it follows by mathematical induction
that $\forall n \in \text{Nat} . P(n)$. Thus all horses are the same colour.   □

Exercise 8. What is the flaw in the proof given above? (Please try to work
this out yourself, and then check the answer in the Appendix.)

7.9   Limitations of Induction

Induction can be used to prove that every element of a set satisfies a certain
property. The set may be finite or infinite. When the theorem states a property of an arbitrary natural number, or an arbitrary list, the inductive proof
establishes that an infinite number of values satisfy the theorem.

Nevertheless, there are some limits on what can be proved using mathematical induction. One such limit is that if the set is infinite, it must be countable
(that is, it must be possible to enumerate its elements, so that each one is associated with a unique natural number). Another limitation, which is particularly
important for computing applications, is that ordinary induction cannot prove

properties of infinite objects; it just proves properties of an infinite number of finite objects, which is not the same thing at all!

An example will clarify this issue. Suppose we define a function

$$reverse \; :: \; [a] \rightarrow [a]$$

which takes a list and returns a new list with the same elements, but in reverse order. For example, $reverse \; [1, 2, 3] \; = \; [3, 2, 1]$. Now, we want to state a theorem which says that if we reverse a list twice, we get the same list back. The following equation is one attempt to say that:

$$reverse \; (reverse \; xs) \; = \; xs$$

Alternatively, we might use extensional equality of functions, and just write

$$reverse \, . \, reverse \; = \; id.$$

It is straightforward to prove the first equation using induction, and the second equation follows immediately using the extensional definition of functional equality.

Theorem 62. $reverse \, . \, reverse = id$

Unfortunately, this theorem is untrue!

To see the problem, let's consider a concrete example. We will choose xs to be $[1..]$, which is the Haskell notation for the infinite list $[1, 2, 3, \ldots]$. Now consider the following two expressions:

- $head \; (reverse \; (reverse \; [1..]))$

- $head \; [1..]$

Now the first of these expressions will go into an infinite loop, because the second (outermost) application of reverse needs to find the last element of its argument before it can return anything, and it will never find the last element of an infinite list. The second expression, however, does *not* go into an infinite loop; it returns the result 1 immediately. Yet, according to the dubious equations stated above, we should be able to replace $reverse \; (reverse \; [1..])$ by $[1..]$ without changing the value! What has gone wrong?

The problem is that mathematical induction only establishes that the theorem is valid for every element of the chain that is connected to the base case by a finite number of steps. *It does not establish that the theorem is true for infinite lists, which are not reachable from the base case in a finite number of steps.* This means that all of our theorems over lists that were proved using induction have actually been proved only for finite lists. Note that there are an infinite number of lists of finite length; thus the induction is proving that a property holds for an infinite number of values, but it does not establish whether the property holds for values that are infinite in size.

There is a related point about natural numbers that sometimes confuses people. When we use induction over natural numbers, using 0 as the base case and $P(n) \rightarrow P(n+1)$ for the inductive case, we have established that the theorem holds for every natural number, and there is an infinite number of naturals. However, this does *not* prove that the theorem holds for infinity itself. There is an infinite number of naturals, but infinity is not itself a natural number.

Exercise 9. Check that Theorem 62 holds for the argument $[1, 2, 3]$.

Exercise 10. Prove the following theorem, using induction:

$$reverse\ (xs\!+\!\!+\!ys) = reverse\ ys\!+\!\!+\!reverse\ xs$$

Then decide whether this theorem happens to be true for infinite lists like $[1..]$. Try to give a good argument for your conclusion, but you don't have to prove it.

Exercise 11. Use induction to prove Theorem 62. $reverse\ (reverse\ xs) = xs$

Exercise 12. Give a careful proof that Theorem 62 does not hold for infinite lists.

7.10 Suggestions for Further Reading

Concrete Mathematics, by Graham, Knuth and Patashnik [14] covers the more advanced mathematical techniques used in the analysis of algorithms. They include a number of problems on induction, and also cover in depth the related topic of recurrences.

Many mathematics books contain more advanced examples of induction proofs. An entire chapter is devoted to induction in Engel's book, *Problem-Solving Strategies* [11], which is a good general source book for mathematical problems.

The textbooks on Haskell cited in Chapter 1 give examples of inductive proofs about recursive programs. The Bird-Meertens calculus [4] develops an extensive theory of programming, including many good applications of induction.

7.11 Review Exercises

Exercise 13. (This problem is from [11], where you can find many more.) The nth Fibonacci number is defined as follows:

```
fib :: Integer -> Integer
fib 0 = 0
fib 1 = 1
fib (n+2) = fib n + fib (n+1)
```

The first few numbers in this famous sequence are $0, 1, 1, 2, 3, 5, \ldots$. Prove the following:

$$\sum_{i=1}^{n} \textit{fib } i \;=\; \textit{fib } (n+2) - 1$$

Exercise 14. Recall Theorem 53, which says

$$\textit{sum } (\textit{map } (1+) \; \textit{xs}) \;=\; \textit{length } \textit{xs} \;+\; \textit{sum } \textit{xs}.$$

Explain in English what this theorem says. Using the definitions of the functions involved (*sum*, *length* and *map*), calculate the values of the left and right hand sides of the equation using $\textit{xs} = [1, 2, 3, 4]$.

Exercise 15. Invent a new theorem similar to Theorem 53, where $(1+)$ is replaced by $(k+)$. Test it on one or two small examples. Then prove your theorem.

Exercise 16. Prove $\textit{sum } . \; \textit{map length } = \; \textit{length } . \; \textit{concat}.$

Chapter 8

Relations

There are many kinds of relationship that occur in everyday life. Some of these describe how the members of a family are related to each other: parent, child, brother, sister, sibling. We could also have a relation called *is in* for cities and countries: for example, London is in Great Britain, and Paris is in France. Or we could have a relation that describes which make of car is produced by which manufacturer. Relations are used in mathematics to describe how two numbers are related to each other; for example expressions like $x < y$ and $p \geq q$ use the relations $<$ and $\geq$.

Similar examples abound in computing, and many branches of computer science use the terminology of relations to describe concepts precisely. Relations are naturally at the heart of relational databases; they are used heavily in the description of programming language syntax; they provide a good notation for representing the internal information required for web search engines, and so on.

Since relations are ubiquitous and important, it is useful to define them as mathematical objects, and to describe their properties. In this chapter we will see how to define relations using set theory, and how to perform various calculations with them.

8.1 Binary Relations

A *binary relation* is used to describe the relationship between two objects. The word *binary* here means simply that there are *two* objects involved; it has nothing at all to do with binary number representations, or binary files. General relationships among any number n of objects are called *n-ary relations*, but binary relations are the most important in computing, and we will restrict ourselves to those for the time being.

Definition 27. A *binary relation* R with type $R :: A \times B$ is a subset of $A \times B$, where A is the *domain* and B is the *codomain* of R. For $x \in A$ and $y \in B$, the notation $x\,R\,y$ means $(x, y) \in R$.

Example 17. Let P be the set of people $\{Bill, Sue, Pat\}$, and let A be the set of animals $\{dog, cat\}$. The relation $has :: P \times A$ describes which person has which kind of animal. Suppose that Bill and Sue both have a dog, and Sue and Pat have a cat. We would represent this information by writing the following relational expressions:

$$Bill\ has\ dog \qquad Sue\ has\ dog$$
$$Sue\ has\ cat \qquad Pat\ has\ cat$$

but the statement '*Bill has cat*' is false. Written out in full, the relation is

$$has = \{(Bill, dog), (Sue, dog), (Sue, cat), (Pat, cat)\}.$$

Example 18. Let R be the set of real numbers. Then $(<) \subset (R \times R)$ is the 'less than' relation and consists of the set of all ordered pairs (x, y) such that $x < y$. Since $(<)$ has an infinite number of elements, we can write it out only partially; for example,

$$(<) \ = \ \{\ldots, (-35.2, -12.1), (-1, 2.7), \ldots\}.$$

Example 19. Many databases use relations to represent the data, since this is a good way to associate different pieces of information with each other. For example, a relation can be used to specify that a person's name is related to that person's address. These are called *relational databases*.

Suppose that you are building a relational database which maintains genealogical data about the families in a town. One of the relations in your database might be the *IsFatherOf* relation. If *John* is the father of *Mary* and *Peter*, then two pairs in the relation have *John* as their first component: $\{(John, Mary), (John, Peter)\}$. This relation states two facts: *John IsFatherOf Mary* and *John IsFatherOf Peter*.

We could also represent the relationship between *John* and his children using a single tuple: $\{(John, Mary, Peter)\}$. This is not a binary relation; it is a more general form of relation called an *n*-ary relation.

Example 20. What would have happened if the pairs had been written as (*Mary, John*) and (*Peter, John*)? Then they would have had an entirely different meaning, asserting *Mary IsFatherOf John* and *Peter IsFatherOf John*, which is not what was intended. It is important to remember that the pairs in a relation are *ordered* pairs. The pairs (1,2) and (2,1) are not equal to each other, nor are the pairs (1,2) and (1,3). However, (2,1) is equal to (2,1).

Example 21. Let's consider two sets, *Children* and *Adults*. The set *Children* includes *Joe*, *Anne* and *Susan*, and the set *Adults* includes *Ray*, *John* and *Dinah*. We would like to create a relation *SmallFamilies* that pairs each child with every possible adult. This is done by taking each child in the first set and

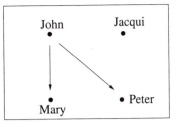

Figure 8.1: The Digraph of the *IsFatherOf* Relation

pairing it in turn with each adult in the second set; that is, we are taking the cross product of the two sets, denoted *Children* × *Adults*:

SmallFamilies =
{(*Joe, Ray*), (*Joe, John*), (*Joe, Dinah*),
(*Anne, Ray*), (*Anne, John*), (*Anne, Dinah*),
(*Susan, Ray*), (*Susan, John*), (*Susan, Dinah*)}

8.2 Representing Relations with Digraphs

Sometimes a diagram provides a good way to visualise a relation. There is a representation of binary relations called a *digraph* which is convenient for computing, and which also is well suited for diagrams. Every element of the domain and codomain in a digraph diagram is represented by a labelled dot (called an *element* or *node*) in the graph, and every pair (x, y) in the relation is represented by an arrow going from x to y (which may be called an *arrow* or *arc*).

Example 22. Figure 8.1 shows a graph illustrating the `IsFatherOf` relation. There is a node for each of *John, Mary, Peter*, and *Jacqui*. There are two arrows, from *John* to *Mary* and *John* to *Peter*.

Many graphs contain some nodes that have no arrows: for example, the node *Jacqui* in Figure 8.1. This means that we need to specify the graph by giving both the set of nodes and the set of arcs.

Some relations have the same set as their domain and codomain, so the relation has a type of the form $R :: A \times A$. In such cases, you draw the graph by writing (and labelling) a dot for every element of A, and then draw in the arrows. Sometimes the domain and codomain are *disjoint*, which means that no element appears in both the domain and codomain. When this happens, it is helpful to keep the dots representing the domain together in the graph diagram, and separate from the dots representing the codomain. Figure 8.2 shows the graph for the relation $R :: A \times A = \{(1, 4), (2, 6)\}$ where $A = \{1, 2, 4, 6\}$.

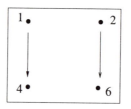

Figure 8.2: The Digraph $(\{1,2,4,6\},\{(1,4),(2,6)\})$

It is important to remember that a relation is more than just a set of ordered pairs; the domain and codomain must be specified, too. In drawing the graph of a relation, you should either draw a dot (or node) for every element of the domain and codomain, and use the layout to indicate exactly what these sets are, or you should specify them explicitly.

Definition 28. Let A be a set, and let R be a binary relation $R :: A \times A$. The *digraph* D of R is the ordered pair $D = (A, R)$.

Example 23. The digraph of the relation $R :: A \times A$, where $R = \{(1,2), (2,3)\}$ and $A = \{1,2,3\}$, is $(\{1,2,3\}, \{(1,2),(2,3)\})$.

Example 24. The digraph of the relation $R :: A \times A$, where $R = \{(1,2), (2,3)\}$ and $A = \{1,2,3,4,5,6\}$, is $(\{1,2,3,4,5,6\}, \{(1,2),(2,3)\})$. Note that this is a different relation than in the previous example, although the set of ordered pairs is identical. The digraph representation records the domain and codomain, giving a precise and complete description of the relation.

This has some interesting implications. For example, two graphs may show an empty relation that contains no arrows (no ordered pairs), but the relations are not equivalent unless their domains and codomains are equal.

Many relations have arcs that are connected to each other in a special way. For example, a set of arcs connected in a sequence is called a (directed) *path*.

Definition 29. A *directed path* is a set of arcs which can be arranged in a sequence, so that the end point of one arc in the sequence is the start point of the next.

Example 25. The sets $\{(1,2), (2,3), (3,4)\}$ and $\{(1,3), (3,1)\}$ are both paths, but the set $\{(1,2), (5,6)\}$ is not.

8.3   Computing with Binary Relations

It is common to compute directly with relations. Throughout this chapter we will use the computer as a calculator for expressions on relations; this is a good way to become accustomed to all the operations on relations since they are used to specify computations formally in many specialised areas of computer science.

A relation $R :: A \times B$, with domain A and codomain B, can be represented as a list of type [(A,B)]. Each element of the list has type (A,B), and is a pair of the form (x,y) where x::A is in the domain and y::B is in the codomain.

Often we impose two restrictions on a relation in order to make it easier to compute with it: (1) there is a finite number of elements in the relation, and (2) the types of the domain and codomain must be in the classes Eq and Show, so that we can compare and print elements of the relation.

Example 26. The relation of colour complements can be represented as follows:

```
data Colour = Red | Blue | Green | Orange | Yellow | Violet
              deriving (Eq, Show)
colourComplement :: Digraph Colour
colourComplement =
  ([Red,Blue,Green,Orange,Yellow,Violet],
   [(Red,Green), (Green,Red),
    (Blue,Orange), (Orange,Blue),
    (Yellow,Violet), (Violet,Yellow)])
```

To say 'the colour complement of red is green', we would write either of the following:

$$\text{Red } colourComplement \text{ Green}$$
$$(Red, Green) \in colourComplement$$

In the example above, we must include both (*Red, Green*) and (*Green, Red*) in *colourComplement*. If we omitted either one of these, we would have a different relation.

The function domain takes a relation and returns its domain:

```
domain ::
  (Eq a, Show a, Eq b, Show b) => Set (a,b) -> Set a
```

For example, domain colourComplement returns the set of colours in the domain of the relation, which is

$$\{Red, Green, Blue, Orange, Yellow, Violet\}.$$

The set is represented as a list, but there is no significance to the order of elements in the list.

The *codomain* function is similar:

```
codomain ::
  (Eq a, Show a, Eq b, Show b) => Set (a,b) -> Set b
```

Many of the operations and functions on sets that we defined in Chapter 4 are useful for working on relations, including `crossproduct` and `setEq`.

Exercise 1. Work out the values of the following expressions, and then check your answer by evaluating the expressions with the computer.

```
domain [(1,100),(2,200),(3,300)]
codomain [(1,100),(2,200),(3,300)]
crossproduct [1,2,3] [4]
```

Exercise 2. The following list comprehensions define a list of ordered pairs. What relations are represented by these lists? Give the domain and the codomain, as well as the pairs themselves.

(a)   `[(a,b) | a <- [1,2],`
                  `b <- [3,4]]`

(b)   `[ (a,b) | a <- [1,2,3],`
                  `b <- [1,2,3],`
                  `a == b]`

(c)   `[ (a,b) | a <- [1,2,3],`
                  `b <- [1,2,3],`
                  `a < b]`

8.4 Properties of Relations

Many relations share interesting and useful properties. For example, we know that if person a is a sibling of person b and b is a sibling of c, then a is also a sibling of c. In a similar way, if x, y and z are numbers, and we know that $x < y$ and $y < z$, then it must also be the case that $x < z$. These two examples show that the *sibling* relation and the $(<)$ relation have essentially the same property (which is called 'transitivity'). In this section we define a variety of such relational properties.

8.4.1 Reflexive Relations

In a *reflexive relation*, every element of the domain is related to itself.

Definition 30. A binary relation R over A is *reflexive* if xRx for every element x of the domain A.

When a reflexive relation is shown in a graph diagram, there must be an arrow from every dot in the domain back to itself.

Example 27. The relation $R :: A \times A$, where $A = \{1, 2\}$ and

$$R = \{(1, 1), (1, 2), (2, 2)\},$$

is reflexive.

Example 28. Let $R :: A \times A$ be a relation, where $A = \{1, 2, 3\}$ and $R = \{(1, 1), (2, 2)\}$. R is not reflexive, but if we added $(3, 3)$ to R then it would be reflexive.

Example 29. A relation *SameFamily*, such that x *SameFamily* y for any two people x and y who are in the same family, is reflexive (because you are in the same family as yourself).

Example 30. The following relations on numbers are reflexive: equality $(=)$, greater than or equal $(\geq)$ and less than or equal $(\leq)$.

Example 31. The following relations on numbers are not reflexive: inequality $(\neq)$, less than $(<)$ and greater than $(>)$.

The Haskell software tools include many functions that test relations for their properties. The function `isReflexive` returns a Boolean which is `True` if the relation is reflexive:

```
isReflexive ::
   (Eq a, Show a) => Digraph a -> Bool
```

Example 32. Consider the following digraphs:

```
a = [1,2,3]
digraph1 = (a,[(1,1), (1,2), (2,2), (2,3), (3,3)])
digraph2 = (a,[(1,2), (2,3), (3,1)])
digraph3 = (a,[(1,1), (1,2), (2,2), (2,3)])
```

The first one, in `digraph1`, is reflexive, since a contains 1, 2 and 3, and all the pairs $(1, 1)$, $(2, 2)$ and $(3, 3)$ appear in `digraph1`. However, 2 is not reflexive because $(1, 1)$ doesn't appear in its set of ordered pairs. (An equally good argument is that $(2, 2)$ doesn't appear, or $(3, 3)$ doesn't appear—all you have to do to show that a relation is not reflexive is to show that there is *some* element x of the domain where (x, x) doesn't appear in the set of ordered pairs.) Finally, `digraph3` is also not reflexive, because $(3, 3)$ is not in the set of ordered pairs.

8.4.2 Irreflexive Relations

A relation is *irreflexive* if no element of its domain is related to itself.

Definition 31. A binary relation R over A is *irreflexive* if, for every $x \in A$, it is *not* the case that xRx.

Example 33. The greater than $(<)$ and less than $(>)$ relations over numbers are irreflexive, since $x < x$ and $x > x$ are always false.

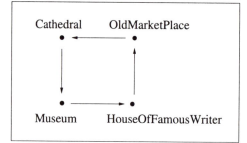

Figure 8.3: The *ByBus* Relation

As long as the domain A of a relation $R :: A \times A$ is non-empty, then it is impossible for R to be both reflexive and irreflexive. To see this, consider some element x of the domain (such an x must exist, since the domain is not empty). If R is reflexive then $(x, x) \in R$, but if R is irreflexive then $(x, x) \notin R$, and both cannot be true.

Example 34. The empty relation $R :: \phi \times \phi$ is reflexive and also irreflexive. In both cases the conditions are met vacuously.

Example 35. Many relations among people are irreflexive. For example, the relations *IsMarriedTo* and *IsChildOf* are irreflexive relations, because no one can marry themselves, or be their own child.

It often happens that a relation is not reflexive and it is also not irreflexive. For example, let $A = \{1, 2, 3, 4, 5\}$ be the domain and codomain of the relation $R = \{(1, 3), (2, 4), (3, 3), (3, 5)\}$. Then R is not reflexive (for example, $(1, 1)$ is not in R) but it is also not irreflexive (because $(3, 3)$ *is* in R).

Suppose that we are visiting a city in France, and want to see several buildings by bus. We can get a bus schedule and look at it, note down the buildings and draw an arrow between each pair of nodes that the bus will visit (Figure 8.3).

```
{(Cathedral, Museum),(Museum, HouseOfFamousWriter),
 (HouseOfFamousWriter, OldMarketPlace),
 (OldMarketPlace, Cathedral)}
```

The bus is travelling in a *cycle*, a path that starts and stops at the same node. However, the ByBus relation is not reflexive: the bus isn't going to waste time cycling around one place and returning to it without going anywhere else. The ByBus relation is *irreflexive*.

The following Haskell function determines whether a binary relation is irreflexive:

```
isIrreflexive ::
  (Eq a, Show a) => Digraph a -> Bool
```

Exercise 3. For each of the following Digraph representations of a relation, draw a graph of the relation, work out whether it is reflexive and whether it is irreflexive, and then check your conclusion using the isReflexive and isIrreflexive functions:

```
([1,2,3],[(1,2)])
([1,2],[(1,2),(2,2),(1,1)])
([1,2],[(2,1)])
([1,2,3],[(1,2),(1,1)])
```

Exercise 4. Determine whether each of the following relations on real numbers is reflexive and whether it is irreflexive. Justify your conclusions.

(a) less than ($<$)

(b) less than or equal to ($\leq$)

(c) greater than ($>$)

(d) greater than or equal to ($\geq$)

(e) equal ($=$)

(f) not equal ($\neq$)

We can't use the isReflexive and isIrreflexive functions on relations with an infinite domain. However, if we restrict the domain to the natural numbers from 0 through 100, it's possible to represent the relations completely and check them with the software tools. The following binary relation representations, with domain $N100 = \{0, \ldots, 100\}$, are defined in the software tools:

```
lessThan_N100,    lessThanOrEq_N100,
greaterThan_N100, greaterThanOrEq_N100,
equals_N100,      notEq_N100
    :: Digraph Int
```

Using these finite relations, use the computer to check your results. Note that a partial check like this does *not* prove anything about the infinite relations, but it is guaranteed to give the correct result for a finite relation on the first hundred natural numbers.

8.4.3 Symmetric Relations

Some relations have the property that the order of two related objects does not matter; that is, if xRy it must also be true that yRx. Such a relation is called a *symmetric* relation.

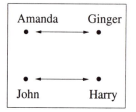

Figure 8.4: The *IsSiblingOf* Relation

Definition 32. Let $R :: A \times A$ be a binary relation. Then R is *symmetric* if $\forall x, y \in A.\ xRy \to yRx$.

Example 36. Equality on real numbers $(=)$ is symmetric, because if $x = y$ then also $y = x$. The equality relation is commonly defined for sets, and it is always symmetric; in fact, one of the essential properties of an abstract equality relation is that it must be symmetric.

Example 37. The family relation *IsSiblingOf* is symmetric.

Example 38. The family relations *IsBrotherOf* and *IsSisterOf* are not symmetric: for example, the term 'Robert *IsBrotherOf* Mary' is true, but 'Mary *IsBrotherOf* Robert' is false.

When you draw the graph diagram for a symmetric relation, every arc from a to b will have a matching arc from b back to a. The notation can be simplified by putting an arrowhead on both sides of every arc.

Example 39. Here is a possible definition of the `IsSiblingOf` relation, shown in Figure 8.4:

```
{(John, Harry), (Harry, John),
 (Amanda, Ginger), (Ginger, Amanda)}
```

Example 40. The relation $R = \{(1,2),(2,1),(2,3),(3,2)\}$ is symmetric.

Example 41. The relation $R = \{(1,2),(1,3),(3,1)\}$ is not symmetric, because $(1,2) \in R$ but $(2,1) \notin R$.

Exercise 5. Is the family relation *IsChildOf* symmetric?

Exercise 6. Suppose a relation $R :: A \times A$, where A is non-empty and reflexive, but it has *only* the arcs required in order to be reflexive. Is R symmetric?

Exercise 7. In the definition of a symmetric relation, can the variables x and y can be instantiated by a single node?

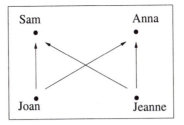

Figure 8.5: The *IsChildOf* Relation

8.4.4 Antisymmetric Relations

An *antisymmetric* relation is one where for all distinct values a and b, it is *never* the case that both aRb and bRa.

Example 42. The less-than relation ($<$) is antisymmetric, since it cannot be true that $x < y$ and also $y < x$.

Example 43. The family relation *IsChildOf* is antisymmetric; if x is a child of y, than y must be the *parent*—not the child—of x. For example, suppose the *IsChildOf* relation contains the following ordered pairs (Figure 8.5):

$$\{(Joan, Sam), (Jeanne, Sam), (Joan, Anna), (Jeanne, Anna)\}$$

Notice that this relation never has both a pair (x, y) and also a pair (y, x).

The antisymmetric property is defined formally as follows:

Definition 33. A binary relation $R :: A \times A$ is *antisymmetric* if

$$\forall x, y \in A. \; xRy \wedge yRx \rightarrow x = y.$$

The graph of an antisymmetric relation may contain some cycles; for example the relation $R = \{(1, 2), (2, 3), (3, 1)\}$ has a cycle from 1 to 2 to 3 and back to 1, and the relation $R_2 = \{(1, 1)\}$ has a trivial cycle containing just 1. However, if an antisymmetric relation does have a cycle, then the length of the cycle cannot be 2, although it may be 1, or greater than 2. In other words, this graph will have no cycles of length 2, but it can have cycles of any other length.

Example 44. Given the set $A = \{1, 2, 3\}$, the relation

$$R :: A \times A = \{(1, 2), (2, 1), (2, 3), (3, 1)\}$$

is not anti-symmetric because both $(1, 2)$ and $(2, 1)$ appear in the set of ordered pairs.

Example 45. Given the set $A = \{1, 2, 3\}$, the relation

$$R :: A \times A = \{(1,1), (1,2), (2,3), (3,1)\}$$

is anti-symmetric.

Example 46. Given the set $A = \{1, 2, 3, 4\}$, and $R_1, R_2, R_3 :: A \times A$, the relations

$$R_1 = \{(1,2), (2,3), (4,1)\}$$

and

$$R_2 = \{(1,1), (2,2)\}$$

are both antisymmetric, but

$$R_3 = \{(1,3), (3,1), (2,3), (3,2)\}$$

is not antisymmetric.

If a relation $R :: A \times A$ is antisymmetric, both of the following statements must be true:

$$\forall x, y \in A.\ \ x \neq y\ \rightarrow\ \neg(xRy \wedge yRx)$$
$$\forall x, y \in A.\ \ x \neq y\ \rightarrow\ \neg xRy \vee \neg yRx$$

Both propositions say that for two distinct elements of the domain, the graph diagram of R contains at most one arrow connecting them.

Example 47. Suppose that we were misanthropic and thought people didn't treat each other well in general. When told that a *Helps* b and b *Helps* a, we might retort that a and b must therefore be the same person! We could express this gloomy view of the world as

$$\forall x, y \in WorldPopulation.\ \ x\ Helps\ y\ \wedge\ y\ Helps\ x\ \rightarrow\ x = y.$$

The software tools define the following functions, which determine whether a finite binary relation is symmetric or antisymmetric:

```
isSymmetric, isAntisymmetric ::
  (Eq a, Show a) => Digraph a -> Bool
```

Exercise 8. First work out whether the relations in the following expressions are symmetric and whether they are antisymmetric, and then check your conclusions by evaluating the expressions with Haskell:

```
isSymmetric ([1,2,3],[(1,2),(2,3)])
isSymmetric ([1,2],[(2,2),(1,1)])
isAntisymmetric ([1,2,3],[(2,1),(1,2)])
isAntisymmetric ([1,2,3],[(1,2),(2,3),(3,1)])
```

Exercise 9. Which of the following relations are symmetric? Antisymmetric?

(a) The empty binary relation over a set with four nodes;

(b) The $=$ relation;

(c) The $\leq$ relation;

(d) The $<$ relation.

8.4.5 Transitive Relations

If x, y and z are three people, and you know that x is a sister of y and y is a sister of z, then x must also be a sister of z. Similarly, if you know that $x < y$ and also that $y < z$, it must also be the case that $x < z$. Relations that have this property are called *transitive* relations.

Definition 34. A binary relation $R :: A \times B$ is *transitive* if

$$\forall x, y, z \in A.\ xRy \wedge yRz \rightarrow xRz.$$

Example 48. The relation $R = \{(1,2), (2,3), (1,3)\}$ is transitive because it contains $(1,3)$, which is required by the presence of $(1,2)$ and $(2,3)$.

Example 49. The relation $R = \{(1,2), (2,3)\}$ is not transitive because there are pairs $(1,2)$ and $(2,3)$ but there is no pair $(1,3)$.

The $(=)$ relation is transitive, as is the *IsAncestorOf* relation.

Example 50. Perhaps surprisingly, the *IsMarriedTo* relation is not transitive. It is certainly symmetric, since if x *IsMarriedTo* y then it must also be the case that y *IsMarriedTo* x. Suppose, however, that x and y are two married people. Then (x, y) and (y, x) are both in the relation, so, if it were transitive, then (x, x) would also need to be in the relation. Nobody is married to themselves, so this cannot be, and the relation is not transitive.

Example 51. Suppose we are flying from one city to another. The relation *FlightTo* describes the point-to-point flights that are available: for example, $(London, Paris) \in FlightTo$ because there is a direct flight from London to Paris. This relation is not transitive, because there are flights from many small cities to London, but those small cities don't have direct flights to Paris. However, the *ReachableByAir* relation *is* transitive. In effect, the airlines define the *FlightTo* relation, and the travel agents extend this to the more general *ReachableByAir* relation, which may involve several connecting flights.

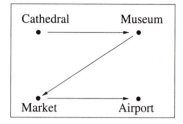

Figure 8.6: The *CityMap* Relation

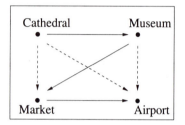

Figure 8.7: The Transitive *CityMap* Relation

As the previous example suggests, a binary relation R can be extended to make a new binary relation R^T, such that $R \subseteq R^T$ and R^T is transitive. This often entails adding several new ordered pairs. For example, suppose we have a relation *CityMap* which defines direct street connections, so that $(x, y) \in$ *CityMap* if there is a street connecting x directly with y (Figure 8.6). The relation could be defined (for a small city) as

$$\{(Cathedral, Museum), (Museum, Market), (Market, Airport)\}.$$

The *CityMap* relation is not transitive, because there is a street path from *Cathedral* to *Market*, but no street connects them directly. Just adding the pair (*Cathedral, Market*) is not enough to make the relation transitive; a total of three ordered pairs must be added. These are shown as dashed arrows in Figure 8.7. The new pairs that we added to the relation are

$$\{(Cathedral, Market), (Cathedral, Airport), (Museum, Airport)\}.$$

A transitive relation provides a short cut for every path of length 2 or more. To make a relation transitive, we must continue adding new pairs until the new relation is transitive. This process is called taking the *transitive closure* of the relation.

The software tools contain a definition of the following function, which determines whether a finite binary relation is transitive:

```
isTransitive ::
 (Eq a, Show a) => Digraph a -> Bool
```

Exercise 10. Determine by hand whether the following relations are transitive, and then check your conclusion using the computer:

```
isTransitive ([1,2],[(1,2),(2,1),(2,2)])
isTransitive ([1,2,3],[(1,2)])
```

Exercise 11. Determine which of the following relations on real numbers are transitive: $(=)$, $(\neq)$, $(<)$, $(\leq)$, $(>)$, $(\geq)$.

Exercise 12. Which of the following relations are transitive?

(a) The empty relation;

(b) The *IsSiblingOf* relation;

(c) An irreflexive relation;

(d) The *IsAncestorOf* relation.

8.5 Relational Composition

We can think of a relation $R :: A \times B$ as taking us from a point $x \in A$ to a point $y \in B$, assuming that $(x, y) \in R$. Now suppose there is another relation $S :: B \times C$, and suppose that $(y, z) \in S$, where $z \in C$. Using first R and then S, we get from x to z, via the intermediate point y.

We could define a new relation that describes the effect of doing first R and then S. This is called the *composition* of R and S, and the notation for it is $R; S$.

Example 52. Suppose that we have a relation *Flight* over the set *City*, where $(a, b) \in$ *Flight* if there is an airline flight from a to b. There is also a relation *BusTrip* over *City*, and $(c, d) \in$ *BusTrip* if there is a bus connection from c to d. Now, we are interested in a relation that describes where we can go, starting from a city with an airport. The relation *Flight*; *BusTrip* consists of the set of pairs (x, y) such that you can get from x to y by flying first to some intermediate city y, and then taking the bus from y on to z.

The use of a semicolon (;) as the operator for relational composition is common, but not completely standard. Many older mathematics books omit the relational composition operator, using RS to mean the relational composition of R and S. Computer scientists often prefer to make all operators explicit. The use of a semicolon is intended to suggest sequencing: just as *statement1 ; statement2* in an imperative programming language means 'First execute *statement1* and then execute *statement2*', the relational composition $R; S$ means 'First apply the relation R and then apply the relation S'.

Relational composition is defined formally as follows:

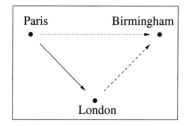

Figure 8.8: The *Route1; Route1* Relation

Definition 35. Let $R_1 :: A \times B$ be a relation from set A to set B, and $R_2 ::$ $B \times C$ be a relation from set B to set C. Their *relational composition* is defined as follows:

$$R_1; R_2 \quad :: \quad A \times C$$
$$R_1; R_2 \quad = \quad \{(a, c) \mid a \in A \wedge c \in C \wedge (\exists b \in B. \; (a, b) \in R_1 \wedge (b, c) \in R_2)\}$$

The definition just says formally that $R_1; R_2$ consists of all the pairs (a, c), such that there is an intermediate connecting point b. This means that $(a, b) \in R_1$ and $(b, c) \in R_2$.

Example 53. When we compose two relations, any two links between a and b in the first relation and b and c in the second produce a new link between a and c. Suppose we have a relation **Route1** linking Paris and London and **Route2** linking London and Birmingham. The composition of **Route1** and **Route2** yields a new route relation which shows that it is possible to travel taking first **Route1** and then **Route2**, starting from Paris and ending at Birmingham (Figure 8.8). In our diagram, the arcs are of three different patterns because they belong to three separate relations.

Sometimes it is useful to compose a relation with itself. A common situation is to start with a relation like *Flight*, which represents trips consisting of just one flight, starting from one city and ending in another one. The composition *Flight; Flight* describes all the trips that can be made using exactly two connecting flights.

Another example arises in databases, where queries often cause the database to derive new information from the facts already available. If the system can predict the requirements of some common queries, then some of this new information can be represented as facts, represented as a new relation, speeding up the execution of future queries.

Suppose that we need to know whether a, who died of a hereditary disease, was a blood relative of b. This could mean calculating all of the descendants of a, then checking to see whether b is among them. It might be better to save some of the work done (space permitting) in calculating the descendants of a,

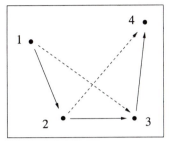

Figure 8.9: The $R_1; R_1$ Relation

so that when we need to know whether a was a blood relative of c, some of the work need not be repeated.

As an example, when determining whether Joseph and Jane are blood relations, we discover that *Joseph IsBloodRelationOf Sarah, Sarah IsBloodRelationOf Jane* and *Jane IsBloodRelationOf Joel.* During this process, we add the newly-discovered fact to the database: *Joseph IsBloodRelationOf Jane.* Now, when we have a query asking whether Joseph is a blood relation of Joel, the new link represents the two links between Joseph and Jane. This reduces the number of links to be traversed.

In creating the composition of two relations, we look for arcs in the first relation that have terminal nodes matching the starting nodes of arcs in the second relation. This operation requires that we systematically check all arcs in R_1 against all arcs in R_2.

Example 54. Let's calculate a relational composition by hand. Let

$$R_1 = \{(1,2), (2,3), (3,4)\}.$$

The composition $R_1; R_1$ is worked out by deducing all the ordered pairs that correspond to an application of R_1 followed by an application of R_1.

First we find all the ordered pairs of the form $(1, x)$. R_1 has only one ordered pair starting with 1; this is $(1, 2)$. This means the first application of R_1 goes from 1 to 2, and the $(2, 3)$ pair means that the second application goes to 3. Therefore the composition $R_1; R_1$ should contain a pair $(1, 3)$. Next, consider what happens starting with 2: the $(2, 3)$ pair goes from 2 to 3, and looking at all the available pairs $\{(1, 2), (2, 3), (3, 4)\}$ shows that 3 then goes to 4. Finally, we see what happens when we start with 3: the first application of R_1 goes from 3 to 4, but there is no pair of the form $(4, x)$. This means that there cannot be any pair of the form $(3, x)$ in the composition $R_1; R_1$. The result of all these comparisons is $R_1; R_1 = \{(1, 3), (2, 4)\}$ (Figure 8.9). In our diagram, the new relation is indicated by arrows with dashes.

The calculation in Example 54 is straightforward and tedious—well suited for computers. The software tools define a function `relationalComposition` that implements this calculation: it defines a new relation giving two existing ones, by working out all the ordered pairs in their relational composition.

```
relationalComposition ::
  (Eq a, Show a, Eq b, Show b, Eq c, Show c) =>
    Set (a,b) -> Set (b,c) -> Set (a,c)
```

Exercise 13. First work out by hand the ordered pairs in the following relational compositions, and then check your results using the computer:

```
relationalComposition [(1,2),(2,3)] [(3,4)]
relationalComposition [(1,2)] [(1,3)]
```

Exercise 14. **(a)** Find the composition of the following relations:

$$\{(Alice, \ Bernard), \ (Carol, Daniel)\} \ \text{and} \ \{(Bernard, Carol)\}.$$

(b) $\{(a,b),(aa,bb)\}$ and $\{(b,c),(cc,bb)\}$

(c) $R; R$, where the relation R is defined as

$$R = \{(1,2),(2,3),(3,4),(4,1)\}.$$

(d) $\{(1,2)\}$ and $\{(3,4)\}$

(e) The empty set and any other relation.

8.6 Powers of Relations

As we saw in the previous section, the composition *Flight; Flight* defines the relation describing all possible trips that consist of two connected flights. More generally, we might want to define a relation defining all possible trips that consist of n connected flights, where n is a natural number. This is called the n-th power of the relation. For a relation R, the nth power is the composition $R; R; \cdots ; R$, where R appears n times, and its notation is R^n. Notice in particular that $R^2 = R; R$, and $R^1 = R$.

When a relation R is composed with itself n times, producing R^n, a path of length n in R from a to b causes there to be a single link (a, b) in the power relation R^n.

Suppose that we have to calculate the relationships between several people in our database, and that the original facts are these (Figure 8.10):

$$\begin{array}{lll} Andrew & IsParentOf & Beth \\ Beth & IsParentOf & Ian \\ Beth & IsParentOf & Joanna \\ Ian & IsParentOf & William \\ William & IsParentOf & Tina \end{array}$$

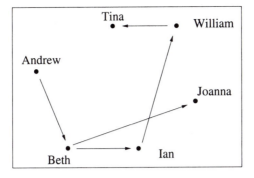

Figure 8.10: The *IsParentOf* Relation

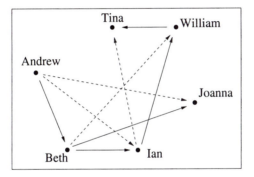

Figure 8.11: The Relation *IsParentOf*²

Now, we will calculate the powers of this relation. The 0'th power is just the identity relation, and the first power *IsParentOf*¹ is simply the *IsParentOf* relation. The higher powers will tell us the grandparents, great grandparents, and great great grandparents. You should expect to see that each of the new relations *IsParentOf*², *IsParentOf*³, and *IsParentOf*⁴ connect up the starting and ending points of a path 2, 3, and 4 arcs long within the original *IsParentOf* relation. In the following diagrams, the arrows with dashes indicate relations defined as a power, while all other arrows belong to the IsParentOf relation.

If we compose the *IsParentOf* relation with itself (i.e. *IsParentOf*²), we have the grandparent relation (Figure 8.11):

Andrew	*IsGrandParentOf*	*Ian*
Andrew	*IsGrandParentOf*	*Joanna*
Beth	*IsGrandParentOf*	*William*
Ian	*IsGrandParentOf*	*Tina*

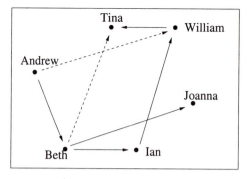

Figure 8.12: The *IsParentOf*^{3} Relation

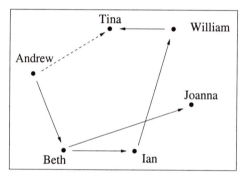

Figure 8.13: The *IsParentOf*^{4} Relation

Now if we compose the *IsGrandParentOf* relation with the original relation, we obtain the *great grand parent* relation (Figure 8.12):

$$Andrew \quad IsGreatGrandParentOf \quad William$$
$$Beth \quad IsGreatGrandParentOf \quad Tina$$

Figure 8.13 shows the composition of the *IsGreatGrandParentOf* relation with *IsParentOf*; thus we have just calculated the fourth power of the *IsParentOf* relation.

$$Andrew \quad IsGreatGreatGrandParentOf \quad Tina$$

We will now give the formal definition of relational powers. The definition is recursive, since we have to define the meaning of R^n for all n. The base case of the recursion will be R^0, which is just the identity relation (it's like taking *zero* flights from a city, which leaves you where you started). The recursive case defines R^{n+1} using the value of R^n.

Definition 36. Let A be a set and let $R :: A \times A$ be a relation defined over A. The nth power of R, denoted R^n, is defined as follows:

$$R^0 = \{(a,a) \mid a \in A\}$$
$$R^{n+1} = R^n; R$$

Example 55. Using the formal definition, we calculate R^4, where

$$R = \{(2,3),(3,2),(3,3)\}.$$

R^0 is just the identity (equality) relation, which contains a reflexive loop for every node. By the definition, $R^1 = R^0; R = R$, since the identity relation composed with R just gives R. The first nontrivial calculation is to find $R^2 = R^1; R = R; R$. We have to take each pair (a, b) in R, and see whether there is a pair (b, c); if so, we need to put the pair (a, c) into R^2. The result of this calculation is $R^2 = \{(2,2),(2,3),(3,2),(3,3)\}$.

Now we have to calculate $R^3 = R^2; R$. We compose

$$\{(2,2),(2,3),(3,2),(3,3)\}$$

with

$$\{(2,3),(3,2),(3,3)\},$$

which yields

$$\{(2,2),(2,3),(3,2),(3,3)\}.$$

At this point, it's helpful to notice that $R^3 = R^2$. In other words, composing R^2 with R just gives R^2 back, and we can do this any number of times. This means that any further powers will be the same as R^2—so we have found R^4 without needing to do lots of calculations with ordered pairs.

Since relational composition is associative, the powers of a relation follow the usual algebraic laws for powers on numbers: for example, $R^{(a+b)} = R^a; R^b$.

A relation whose domain is $\{x_0, \ldots, x_{n-1}\}$ is *cyclic* if it contains a cycle of ordered pairs of the form $(x_0, x_1), (x_1, x_2), (x_2, x_3), (x_3, x_4), \ldots, (x_{n-1}, x_0)$. That is, the relation is cyclic if there is a cycle comprising all the elements of its domain.

Consider what happens to a cyclic relation as we calculate its powers. The relation is defined as

$$R = \{(a,b),(b,c),(c,a)\}.$$

The first power R^1 is just R. The second power R^2 is calculated by working out the ordered pairs in $R; R$; the result is

$$R^2 = \{(a,b),(b,c),(c,a)\}; \{(a,b),(b,c),(c,a)\}$$
$$= \{(a,c),(b,a),(c,b)\}.$$

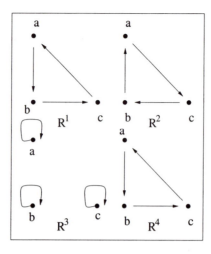

Figure 8.14: Four Powers of a Cyclic Relation

This contains only paths between the start and end points of all the paths of length two in the original relation. Now the third power is

$$R^3 \quad = \quad \{(a,c),(b,a),(c,b)\}; \{(a,b),(b,c),(c,a)\}$$
$$= \quad \{(a,a),(b,b),(c,c)\}.$$

This result contains only arcs connecting the origin and destination points of paths of length *three* in the original relation. What will happen next?

$$R^4 \quad = \quad \{(a,a),(b,b),(c,c)\}; \{(a,b),(b,c),(c,a)\}$$
$$= \quad \{(a,b),(b,c),(c,a)\}$$

Just what we might have expected: each of these arcs represents a path of length 4, so we have started round the cycle again. What can we now say about the powers of this relation? They repeat in a cycle. $R^4 = R^1$, $R^5 = R^2$ and in general $R^{n+3} = R^n$ (Figure 8.14).

The software tools file defines the following function, which takes a set and returns the equality relation on that set.

```
equalityRelation ::
   (Eq a, Show a) => Set a -> Relation a
```

There is also a function that calculates the power of a relation:

```
relationalPower ::
   (Eq a, Show a) => Digraph a -> Int -> Relation a
```

Exercise 15. Work out the values of these expressions, and then evaluate them using the computer:

```
equalityRelation [1,2,3]
equalityRelation ([]::[Int])
```

Exercise 16. Calculate the following relational powers by hand, and then evaluate them using the computer.

```
relationalPower ([1,2,3,4],[(1,2),(2,3),(3,4)]) 1
relationalPower ([1,2,3,4],[(1,2),(2,3),(3,4)]) 2
relationalPower ([1,2,3,4],[(1,2),(2,3),(3,4)]) 3
relationalPower ([1,2,3,4],[(1,2),(2,3),(3,4)]) 4
```

Exercise 17. Why do we not need to check the ordered pairs in R while calculating $R^0; R$?

Exercise 18. Why can we stop calculating powers after finding that two successive powers are the same relation?

Exercise 19. What is R^4 where R is $\{(2,2),(4,4)\}$?

Exercise 20. What is the relationship between adding new ordered pairs to make a relation transitive and taking the power of a relation?

Exercise 21. Suppose a set A contains n elements. How many possible relations with type $R :: A \times A$ are there?

Exercise 22. Given the relation $\{(a,b),(b,c),(c,d),(d,e)\}$, how many times would we have to compose this relation with itself before the empty relation is produced?

Exercise 23. Given the set $A = \{1,2,3\}$ and the relation $R :: A \times A$ where $R = \{(3,1),(1,2),(2,3)\}$, what is the value of R^2? R^3?

8.7 Closure Properties of Relations

In computing applications, we normally want to keep the specification of a relation as small and readable as possible, so that it can be defined and maintained accurately. On the other hand, some computations may require the relation to have some special properties (for example, symmetric or transitive). These special properties would require adding a large number of ordered pairs to the relation, making it harder to maintain. There is a standard technique used in such situations—we define *two* relations:

1. A basic relation, containing just the essential information, is specified;

2. A larger relation is derived from the basic one by adding the ordered pairs required to give it the special properties that are needed.

When circumstances change, only the basic relation is edited by hand. The derived relation is recalculated using a computer.

Example 56. An airline keeps a set of all the flights they offer. This is represented by a relation *Flight*, where $(a, b) \in$ *Flight* if the airline has a direct flight from a to b. However, when a customer asks a question like 'Do you fly from Glasgow to Seattle?', the airline needs a *transitive* relation: if *(Glasgow, New York)* $\in$ *Flight* and also *(New York, Seattle)* $\in$ *Flight*, the answer should be *yes*. Thus the airline's flight-planning staff define the basic relation *Flight*, but the sales staff work with a derived relation which is similar to *Flight*, but which is transitive.

A relation derived in this way is called the *closure* of the basic relation:

Definition 37. The *closure* of a relation R with respect to a given property is the smallest possible relation that contains R and has that property.

Closure is suitable for adding properties that require the *presence* of certain ordered pairs. For example, you can take the symmetric closure of a relation by checking every existing pair (x, y), and adding the pair (y, x) if it isn't already present. However, closure is *not* suitable for properties that require the *absence* of certain ordered pairs. For example, the relation $R = \{(1, 1), (1, 2), (2, 3)\}$ does not have an irreflexive closure, since that would need to contain $(1, 1)$ (since the closure must contain the basic relation), yet it *must not* contain $(1, 1)$ (in order to be irreflexive).

You can give a relation a property such as reflexivity, or transitivity, by creating its reflexive or transitive closure. Notice, however, that the new relation may no longer have all of the properties of the original relation. For example, suppose that a relation is irreflexive, as in $\{(1, 2), (2, 1)\}$. The smallest possible transitive relation containing this one also has the arcs (1,1) and (2,2), which means that it is no longer irreflexive.

8.7.1 Reflexive Closure

The reflexive closure of a relation contains all of the arcs in the relation together with an arc from each node to itself. For example, consider the set

$$\{Red, \, Orange, \, Yellow, \, Green, \, Blue, \, Violet\}.$$

The relation *ContainsPrimaryColour* is given in Figure 8.15 by the solid arcs. It is not reflexive, because the colours that are not primary colours do not contain reflexive arcs. However, the relation *ContainsColour* is reflexive (given by all the arcs in Figure 8.15), and is in fact the reflexive closure of *ContainsPrimaryColour*.

The formal definition that follows defines the reflexive closure of a relation R as the smallest reflexive relation containing the arcs of R.

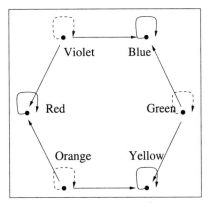

Figure 8.15: The *ContainsColour* and *ContainsPrimaryColour* Relations

Definition 38. Let A be a set, and let $R :: A \times A$ be a binary relation over A. The *reflexive closure* of R is the relation R' such that R' is reflexive, R' is a superset of R, and for any reflexive relation R'', if R'' is a superset of R, then R'' is a superset of R'. The notation $r(R)$ denotes the reflexive closure of R.

Example 57. The reflexive closure of the relation $\{(1, 2), (2, 3)\}$ over the set $\{1, 2, 3\}$ is $\{(1, 1), (2, 2), (3, 3), (1, 2), (2, 3)\}$.

Example 58. The relation $\{(1, 1), (1, 2), (3, 3), (2, 3)\}$ is *not* the reflexive closure of $\{(1, 2), (2, 3)\}$, since it is missing the reflexive arc $(2, 2)$.

The following theorem provides a straightforward method for calculating the reflexive closure of a relation:

Theorem 63. Let A be a set, let E be the equality relation on A, and let R be a binary relation defined over A. Then $r(R) = R \cup E$.

All we have to do to calculate the reflexive closure $R :: A \times A$ is to add self-loops (x, x) for all of the nodes x in the set A over which the relation R is defined.

The software tools file provides the following function, which automates the calculation of reflexive closures:

```
reflexiveClosure ::
    (Eq a, Show a) => Digraph a -> Digraph a
```

Exercise 24. Work out the following reflexive closures by hand, and then check your results using the computer:

```
reflexiveClosure ([1,2,3],[(1,2),(2,3)])
reflexiveClosure ([1,2],[(1,2),(2,1)])
```

Exercise 25. What is the reflexive closure of the relation $R; R$, where R is defined as $\{(1, 2), (2, 1)\}$?

8.7.2   Symmetric Closure

In maintaining a genealogical database, we might enter an ordered pair (a, b) stating that person a *IsMarriedTo* person b, but we don't want to enter another pair saying explicitly that person b *IsMarriedTo* person a, in order to save time typing. However, the *IsMarriedTo* relation should certainly be symmetric. In order to derive this information, the database must calculate the *symmetric closure* of the basic relation that was typed in. It does this by adding only those arcs which are needed to make the relation symmetric.

The formal definition of symmetric closure is very similar to the definition of reflexive closure:[1]

Definition 39. Let A be a set, and let $R :: A \times A$ be a binary relation over A. The *symmetric closure* of R is the relation R' such that R' is symmetric, R' is a superset of R, and for any symmetric relation R'', if R'' is a superset of R, then R'' is a superset of R'. The notation $s(R)$ denotes the symmetric closure of R.

Sometimes it is useful to turn around a relation and use its ordered pairs in reverse. This is called the *converse* of the relation:

Definition 40. Let A and B be sets, and let $R :: A \times B$ be a binary relation from A to B. The *converse* of R, written R^c, is the binary relation from B to A defined as follows:

$$R^c = \{(b, a) | (a, b) \in R\}.$$

This definition says that if you reverse the order of the components in a relation's arcs, you create its converse. The symmetric closure of a relation is the union of the relation and its converse.

The converse operation provides an alternative way to calculate the symmetric closure of a relation. The idea is that we take the original relation R, and add to it the set of reversed ordered pairs, which is R^c. The following theorem states this formally:

Theorem 64. Let A be a set and let $R :: A \times A$ be a binary relation over A. Then the symmetric closure $s(R) = R \cup R^c$.

Example 59. Suppose that we had a relation $\{(1, 1), (2, 2)\}$ and wanted to create its symmetric closure. How many arcs would need to be added? None, because the relation is already symmetric.

Example 60. On the other hand, suppose that we had the relation $\{(1, 2)\}$. It is not symmetric, so when creating its symmetric closure, we add the arc $(2, 1)$ (Figure 8.16).

The following Haskell function calculates the symmetric closure of a binary relation:

---

[1]A useful tip for learning to read mathematics: many definitions follow standard forms, and this makes it easier to read and understand new ones.

Figure 8.16: The Symmetric Closure of $\{(1,2)\}$

```
symmetricClosure ::
  (Eq a, Show a) => Digraph a -> Digraph a
```

Exercise 26. Work out the following symmetric closures by hand, and then calculate them using the computer window:

```
symmetricClosure ([1,2],[(1,1),(1,2)])
symmetricClosure ([1,2,3],[(1,2),(2,3)])
```

Exercise 27. What is the symmetric reflexive closure of the relation

$$\{(a,b),(b,c)\}?$$

Hint: take the reflexive closure first, followed by the symmetric closure of the result.

Exercise 28. Find the reflexive symmetric closure of the relation $\{(a,c)\}$.

8.7.3 Transitive Closure

The transitive closure is one of the most important operations on relations. You can define a relation that describes one step; the transitive closure of the relation then describes the effect of taking n steps, for any n.

Example 61. Suppose that *Flight* is the relation where $(a,b) \in$ *Flight* if there is a direct flight from city a to city b. Then the transitive closure of *Flight* is the relation consisting of pairs (a,b) where b is *reachable* by air from a.

Consider now how to calculate the transitive closure of a relation. As an example, suppose that you need to define the *IsDescendantOf* relation in a database of people. The database contains records for *Zoe, Bruce, Gina, Annabel, Dirk, Kay* and *Don*. We start with the *IsChildOf* relation, defined as follows:

$$\{(Zoe, Bruce), (Gina, Zoe),$$
$$(Bruce, Annabel), (Dirk, Kay),$$
$$(Kay, Don), (Annabel, Kay)\}$$

Observe that the relation *IsDescendantOf* should have all these arcs, and many more. For example, *Gina* is a descendant of *Bruce*.

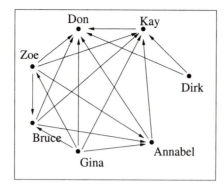

Figure 8.17: The *IsDescendantOf* Relation

How many more arcs need to be added? There should be an arc between the origin and destination points of each path of length 2 or more in the *IsChildOf* relation. In other words, we need the transitive closure of *IsChildOf*.

The powers of the *IsChildOf* relation are as follows:

$$
\begin{aligned}
Is\_Child\_Of^2 &= \{(Zoe, Annabel), (Gina, Bruce), \\
&\quad (Bruce, Kay), (Dirk, Don), (Annabel, Don)\} \\
Is\_Child\_Of^3 &= \{(Zoe, Kay), (Gina, Annabel), (Bruce, Don)\} \\
Is\_Child\_Of^4 &= \{(Zoe, Don), (Gina, Kay)\} \\
Is\_Child\_Of^5 &= \{(Gina, Don)\}
\end{aligned}
$$

The *IsDescendantOf* relation contains the union of all of these powers (Figure 8.17):

$$
\begin{aligned}
IsDescendantOf = \\
\{(Zoe, Bruce), (Gina, Zoe), (Bruce, Annabel), (Dirk, Kay), \\
(Kay, Don), (Annabel, Kay), (Zoe, Annabel), (Gina, Bruce), \\
(Bruce, Kay), (Dirk, Don), (Annabel, Don), (Zoe, Kay), \\
(Gina, Annabel), (Bruce, Don), (Zoe, Don), (Gina, Kay), \\
(Gina, Don)\}
\end{aligned}
$$

The calculations we have just gone through suggest an algorithm for calculating the transitive closure of any relation: calculate all the powers R^1, R^2 and so on, up to R^n, and take their union. There are n nodes in the digraph, so the longest possible path (ignoring cycles) must be no more than $n - 1$ elements long. The transitive closure must provide a short cut for each path, which is why we must include a power of the relation for each possible path length. This leads also to a way to define the transitive closure formally:

Definition 41. Let A be a set of n elements, and let $R :: A \times A$ be a binary relation over A. The *transitive closure* of R is defined as

$$t(R) = \bigcup_{i=1}^{n} R^i.$$

For example, if a set A has four elements, then the transitive closure of a relation $R :: A \times A$ would be

$$R^1 \cup R^2 \cup R^3 \cup R^4.$$

The *IsDescendantOf* relation is the union of as many powers of the *IsChildOf* relation as there are people in the *IsChildOf* relation's domain.

Example 62. Using the definition, we calculate the transitive closure of $R = \{(1,2), (2,3), (3,2), (3,4), (4,4)\}$. There are four elements in the set over which the relation is defined (we haven't specified otherwise), so we shall need to calculate the union of the relations R, R^2, R^3 and R^4 (Figure 8.18). First, we calculate:

$$
\begin{aligned}
R^2 &= \{(1,3), (2,2), (2,4), (3,3), (3,4), (4,4)\} \\
R^3 &= \{(1,2), (1,4), (2,3), (2,4), (3,2), (3,4), (4,4)\} \\
R^4 &= \{(1,3), (1,4), (2,2), (2,4), (3,3), (3,4), (4,4)\}
\end{aligned}
$$

The union of all of these relations is the transitive closure of R:

$$\{(1,2), (1,3), (1,4), (2,2), (2,3), (2,4), (3,2), (3,3), (3,4), (4,4)\}$$

The following function calculates the transitive closure of a relation, using the definition:

```
transitiveClosure ::
   (Eq a, Show a) => Digraph a -> Digraph a
```

Exercise 29. Work out the following transitive closures by hand, and then evaluate them using the computer:

```
transitiveClosure ([1,2,3],[(1,2),(2,3)])
transitiveClosure ([1,2,3],[(1,2),(2,1)])
```

Exercise 30. Given a digraph $(\{1,2,3,4\}, \{(1,2)\})$, what can we do to speed up the transitive closure algorithm, which requires that we take as many powers of this relation as there are nodes in the digraph?

Exercise 31. Find the transitive symmetric closure and the symmetric transitive closure of the following relations:

(a) $\{(a,b), (a,c)\}$

(b) $\{(a,b)\}$

(c) $\{(1,1), (1,2), (1,3), (2,3)\}$

(d) $\{(1,2), (2,1), (1,3)\}$

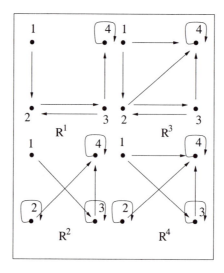

Figure 8.18: The Powers of $\{(1,2),(2,3),(3,2),(3,4),(4,4)\}$

8.8 Order Relations

An *order relation* specifies an ordering which can be used to create a sequence from the elements of its domain. Order relations are extremely important in computing, because data values often need to be placed in a well-defined sequence for processing. The standard mathematical relations less-than ($<$) and less-than-or-equal ($\leq$) are examples of order relations, but there are many more.

One of the most fundamental properties of an order relation is transitivity: if a precedes b in the ordering, and b precedes c, then we surely want a also to precede c. However, some of the other properties of relations may be present or absent, so there are several different kinds of order relation. We will examine these in turn, starting with partial orders.

8.8.1 Partial Order

A *partial order* puts at least some of the elements in its domain into sequence, but not necessarily all of them. There could also be several sequences within a partial order, without any ordering between elements belonging to different subsequences.

Example 63. Suppose that a database of people contains records that specify the breed of dog owned—for those people who have a dog. The records of dog owners could be ordered alphabetically using the breed name, producing a sequence of dog owners. However, this ordering would not include the people

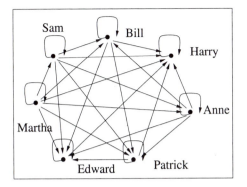

Figure 8.19: The *IsYoungerOrSameAgeAs* Partial Order

who don't have dogs, so it is only a *partial* order. Of course, it might happen that everyone (or no one) owns a dog, in which case we would still technically have a partial order. That is, it is possible that the entire partial order is sorted using some ordering; the point is just that this is not required.

Example 64. Consider the problem of ordering all the records in the database by people's names. Some names are common, so there might be more than one record per name. Therefore this is a partial order.

Example 65. We are programming with a data structure that contains ordered pairs (x, y), and we define an ordering such that the pair (x_1, y_1) precedes the pair (x_2, y_2) if $x_1 \leq x_2 \wedge y_1 \leq y_2$. This is a partial order, because it doesn't specify the ordering between $(1, 4)$ and $(2, 3)$.

The formal definition of partial orders is stated using the properties of relations that we have already defined:

Definition 42. A binary relation R over a set A is a *partial order* if it is reflexive, antisymmetric and transitive.

Example 66. Suppose that we used age to order our database records. Our *IsYoungerOrSameAgeAs* relation is reflexive, antisymmetric, and transitive, so it is a partial order. Figure 8.19 gives its digraph.

Poset diagrams

The purpose of drawing the graph diagram for a relation is to make it easier to understand. It often defeats the purpose to include all of the relation's arcs in the diagram, since there are so many of them. For a partial order, there is no point in drawing the reflexive and transitive arcs, because we know they must be there anyway and they clutter the diagram and make it hard to see the important arcs that tell us about the ordering.

Figure 8.20: A Poset Diagram of the *IsYoungerOrSameAgeAs* Relation

A *poset* (partially-ordered set) diagram is a relation diagram for partial orders, where the distracting transitive and reflexive arcs are omitted. It is important to state explicitly that the diagram shows a partial order (or a poset); without knowing this fact, a reader would not know that the relation also contains the reflexive and transitive arcs.

When drawing a poset diagram, we position it so that all of the arcs point upwards. All of the arcs of the partial order that are *not* implied by the reflexive or transitive properties must be drawn explicitly. We remove the reflexive and transitive arcs, and the arrowheads of the remaining arcs.

If there is a directed path between two nodes in a poset diagram, then those nodes are comparable. An element of a partial order may be comparable to some of the elements, but not to the others.

Example 67. We redraw Figure 8.19 so that it is a poset diagram (Figure 8.20). Now, it is easy to see the record ordering.

Weakest and greatest elements of a poset

The following definitions give the standard terminology used to describe how two elements of a partial order are related to each other:

Definition 43. If there is a directed path from x to y in a partial order (i.e. if x precedes y in the partial order), then x *is weaker than* y. The mathematical notation for this $x \sqsubseteq y$. If $x \sqsubseteq y$ is false, then we write $x \not\sqsubseteq y$.

Definition 44. Two nodes x and y in a partial order are *incomparable* if $x \not\sqsubseteq y \wedge y \not\sqsubseteq x$. That is, x and y are incomparable if there is no directed path from x to y, and there is also no directed path from y to x.

In a finite set of numbers, there must be a unique smallest element and a unique greatest element. However, a poset might have *several* least elements. For example, if x and y are incomparable, but they are both weaker than all

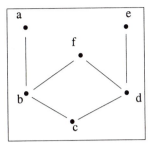

Figure 8.21: A Poset Diagram

the other elements of the poset, then both are least elements. Similarly, there may be several greatest elements. The following definitions define the sets of least and greatest elements formally:

Definition 45. The set of *least elements* of a poset P is

$$\{x \in P \mid \forall y \in P. \ (x \sqsubseteq y \vee (x \not\sqsubseteq y \wedge y \not\sqsubseteq x))\}.$$

That is, the least elements of P are the elements that are either incomparable to or weaker than any other element.

Definition 46. The set of *greatest elements* of a poset P is

$$\{x \in P \mid \forall y \in P. \ (y \sqsubseteq x \vee (x \not\sqsubseteq y \wedge y \not\sqsubseteq x))\}.$$

That is, the greatest elements of P are the elements that are either incomparable to or greater than any other element.

Example 68. Suppose that a family in our database has the following children: *Ray*, aged 17, *Tom*, aged 6 and the twins *Belle* and *Eunice*, aged 5. We can define a partial order ($\sqsubseteq$) based on age, such that $x \sqsubseteq y$ if x is younger than or the same age as y. Even though there are twins, there are no incomparable elements of this poset, since *Belle* $\sqsubseteq$ *Eunice* and also *Eunice* $\sqsubseteq$ *Belle*. However, both of the twins satisfy the requirements for the least element. The set of greatest nodes is {*Ray*} and the set of weakest nodes is {*Belle, Eunice*}.

Example 69. Figure 8.21 shows a poset where the set of weakest elements is {c}, and the set of greatest elements is {a, f, e}.

The following Haskell function, defined in the software tools file, takes a digraph and returns True if the digraph represents a partial order and False otherwise.

```
isPartialOrder ::
  (Eq a, Show a) => Digraph a -> Bool
```

The following two functions each take a relation and an element. The first one returns True if the second argument is a least element in the relation, and False otherwise. The second function returns True if the element is a greatest element in the relation and False otherwise.

```
isWeakest, isGreatest ::
  (Eq a, Show a) => Relation a -> a -> Bool
```

These functions each take a digraph; the first function returns the set of weakest elements while the second function returns the set of greatest elements:

```
weakestSet ::  (Eq a, Show a) => Digraph a -> Set a
greatestSet :: (Eq a, Show a) => Digraph a -> Set a
```

Exercise 32. Work out by hand whether the following digraphs are partial orders, and then check your results using the computer:

```
isPartialOrder ([1,2,3], [(1,2),(2,3)])
isPartialOrder
  ([1,2,3], [(1,2),(2,3),(1,3),(1,1),(2,2),(3,3)])
```

Exercise 33. Calculate the following by hand, and then evaluate using the computer:

```
isWeakest [(1,2),(2,3),(1,3),(1,1),(2,2),(3,3)] 2
isWeakest [(1,2),(1,3),(1,1),(2,2),(3,3)] 3

isGreatest [(1,2),(2,3),(1,3),(1,1),(2,2),(3,3)] 3
isGreatest [(1,2),(1,3),(1,1),(2,2),(3,3)] 1
```

Exercise 34. Calculate the following by hand, and then evaluate using the computer:

```
weakestSet ([1,2,3,4],
            [(1,4),(1,3),(1,2),(1,1),
             (2,3),(2,4),(2,2),(3,4),
             (3,3),(4,4)])
weakestSet ([1,2,3,4],
            [(1,4),(1,2),(1,1),(2,4),
             (2,2),(3,4),(3,3),(4,4)])
greatestSet ([1,2,3,4],
            [(1,2),(3,4),(1,1),(2,2),(3,3),(4,4)])
greatestSet ([1,2,3,4],
            [(2,3),(3,4),(2,4),(1,1),(2,2),(3,3),(4,4)])
```

Exercise 35. What are the greatest and weakest elements in a poset diagram that contains the following arcs:

(a) $\{(a, b), (a, c)\}$

(b) $\{(a, b), (c, d)\}$

(c) $\{(a, b), (a, d), (b, c)\}$

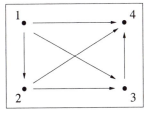

Figure 8.22: A Quasi Order

8.8.2   Quasi Order

A *quasi* order is similar to a partial order, except that it is irreflexive:

Definition 47. A binary relation R over a set A is a *quasi order* if it is ir-reflexive and transitive.

Example 70. The relation $(<)$ on numbers is a quasi order, but $(\leq)$ is not.

Notice that the definition of a quasi order doesn't mention symmetry. Can a quasi order be symmetric? Suppose there are two elements x and y, such that $x \sqsubseteq y$. If the quasi order were symmetric, then we would also have $y \sqsubseteq x$, and since it is also transitive, we then have $x \sqsubseteq x$, which violates the requirement that a quasi order be irreflexive. This argument would not apply, of course, in a trivial quasi order where no two elements are related by $\sqsubseteq$, but non-trivial quasi orders cannot be symmetric.

We should also inquire whether a quasi order can be (or must be) antisymmetric. By definition, it is antisymmetric if $x \sqsubseteq y \wedge y \sqsubseteq x \rightarrow x = y$ for any two elements x and y. Now, if we choose x and y to be the same, then $x \sqsubseteq y \wedge y \sqsubseteq x$ is false, because the quasi order is irreflexive. This means the logical implication is vacuously true. If we choose x and y to be different, then $x \sqsubseteq y \wedge y \sqsubseteq x$ is again false (as we have just shown while discussing symmetry). In all cases, therefore, the definition of antisymmetry is satisfied, but vacuously.

The conclusion is that quasi orders may be symmetric, but only if they are trivial, and they are always antisymmetric, but only because they satisfy the definition vacuously. The properties of symmetry and antisymmetry are uninteresting for quasi orders.

Example 71. Figure 8.22 gives the graph diagram for the quasi order $(<)$ on the set $\{1, 2, 3, 4\}$.

The following function takes a digraph and returns `True` if the relation it represents is a quasi order, and `False` otherwise:

```
isQuasiOrder ::
  (Eq a, Show a) => Digraph a -> Bool
```

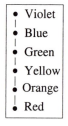

Figure 8.23: A Chain of the Rainbow Colours

Exercise 36. Work out the following expressions, and evaluate them with the computer:

```
isQuasiOrder ([1,2,3,4],[(1,2),(2,3),(3,4)])
isQuasiOrder ([1,2,3,4],[(1,2)])
```

8.8.3 Linear Order

A *linear order* or *total order* is like a partial order, except that it requires that *all* of the relation's elements must be related to each other.

Example 72. The ($\leq$) and ($\geq$) relations on real numbers are total orders: any two numbers x and y can be compared with each other, and it is guaranteed that either $x \leq y$ or $y \leq x$ will be true (and both are true if $x = y$).

Definition 48. A *linear order* is a partial order defined over a set A in which for each element a and b in A, either $a \sqsubseteq b$ or $b \sqsubseteq a$.

Example 73. Suppose that the database recorded the exact time at which each child was born. We could then use a form of $\leq$ to order the children within the families. This information could be useful in a study of the influence of primogeniture on the medical history of an aristocracy.

The elements of a linear order can be said to form a *chain*. When we draw the graph diagram for a chain, we omit the arcs that are implied by transitivity and reflexivity. Without these extra arcs, and because no element can be incomparable to the others, the diagram looks like a real chain. For example, the colours of the rainbow are often given as a chain starting with *Red* and ending with *Violet*. As *Red* light has the longest wavelength and *Violet* the shortest, the relation that imposes this chain ordering on the set of six colours is the $\leq$ relation on the wave frequency (Figure 8.23).

The *isLinearOrder* function takes a digraph and returns `True` if it represents a linear order, and `False` otherwise.

```
isLinearOrder :: Eq a => Digraph a -> Bool
```

Exercise 37. Evaluate the following expressions, by hand and using the computer:

```
isLinearOrder
  ([1,2,3],[(1,2),(2,3),(1,3),(1,1),(2,2),(3,3)])
isLinearOrder
  ([1,2,3],[(1,2),(1,3),(1,1),(2,2),(3,3)])
```

8.8.4 Well Order

A *well order* is a total (or linear) order that has a least element; furthermore, every subset of a well order must have a least element.

The existence of a least element is significant because it provides a base case for recursive functions and for inductive proofs. Note that any total order which has a finite number of elements must have a least element. Some total orders with an infinite number of elements have a least element, and others do not.

Example 74. The ($\leq$) relation on the set $N = \{0, 1, 2, \ldots\}$ of natural numbers is a total order. Furthermore, N has a least element, because $\forall x \in N. \ 0 \leq x$. Therefore ($\leq$) on N is a well order.

Example 75. The ($\leq$) relation on the set $Z = \{\ldots, -2, -1, 0, 1, 2, \ldots\}$ of integers is a total order. However, Z does not have a least element, because

$$\forall x \in Z. \exists y \in Z. (y \leq x \wedge y \neq x).$$

Therefore ($\leq$) on Z is only a total order, and not a well order.

Definition 49. Given a set S and a binary relation R over S, R is a *well order* if R is a linear order and every subset of S that is not empty contains a least element.

Well orders are important because they are countable. Informally, a countable set is a set in which an arbitrary item can eventually be processed by a computer. The set could be infinite: for example, the set of natural numbers is infinite, but *every* element of that set would eventually be reached if we just work on 0, 1, 2, ... in sequence. For example, if a computer started printing the natural numbers, we would eventually see the number 4058000023. However, if it started printing an uncountable set such as the irrational numbers, then it might get stuck printing an infinite number of irrationals *without reaching* the number we are interested in.

Example 76. In our database, each record is given a numeric key that is unique. As there are a finite number of keys in the database, the $\leq$ relation over these keys is a well order.

Exercise 38. We have been watching a computer terminal. Is the order in which people come and use the terminal a total order?

Exercise 39. Is it always possible to count the elements of a linear order?

Exercise 40. Can a set that is not a well order be countable?

8.8.5 Topological Sort

Computers are good at doing one task at a time, in sequence. When an algorithm is working on a data structure, it needs to know which element of the data structure to work on next. Often there is an order relation that must be followed (for example, we might want to output the items in a database in alphabetical order). If we have a total order on the elements of the data structure, the algorithm can use that to find the next piece of work. If, however, we have only a partial order on the data items, then there are several possible orders in which items could be processed while still respecting the order relation. Often we don't care *which* order is used—we just want the algorithm to find one and proceed with the work.

The process of taking a partial order and putting its elements into a total order is called *topological sorting*.

Example 77. Some compilers analyse the order in which procedures call each other. Such a compiler could construct a 'dependency graph' for the program it is translating, where each node corresponds to a procedure, and arcs in the graph correspond to procedure calls. The dependency graph is a partial order. Now, suppose the compiler generates object code for the procedures in the order of their appearance in the call graph, so that the lowest-level procedures are processed first and the highest-level ones are done last, in order to make as many procedure calls as possible into forward references. The compiler uses a topological sort to produce the total order in which it prints the information. The first name in the total order will be a procedure that doesn't call any other procedure, while the last is the top-level procedure with which the program starts execution.

There is a simple and general algorithm for topological sorting. Choose x, one of the elements that is greatest in the set A, and make it the first in the sequence. Now do the same for the set $A - \{x\}$, and continue until A is empty.

Example 78. Suppose that we have a relation that expresses the call graph:

$$\{('A','B'),('B','B'),('B','C'),('B','D'),('C','D')\}$$

What would the topological sort of this graph be? First, the functions that call no other function would appear, followed by the functions that call them, followed by the functions calling *them*, etc. The result would be 'D', 'C', 'B', 'A' (Figure 8.24).

Example 79. What is the topological sort of $\{(1,2),(1,3)\}$ given the nodes $1,2,3$? The sequence $3,2,1$ or $2,3,1$.

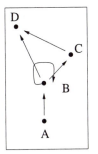

Figure 8.24: A Call Graph

The following function takes a digraph and returns a topological sort of its relation.

```
topsort ::
    (Eq a, Show a) => Digraph a -> Set a
```

Exercise 41. Check to see that the following partial orders are not, in fact, total orders. Use the computer to generate a total order, using a topological sort.

```
topsort ([1,2,3,4],[(1,2),(1,3),(2,3),(1,4),(2,4),
                    (1,1),(2,2),(3,3),(4,4)])
topsort ([1,2,3],[(1,2),(1,3),(1,4),(1,1),(2,2),(3,3)])
```

8.9 Equivalence Relations

Some relations can be used to break a set up into several categories or 'partitions', where each element of the set belongs to just one of the categories. Such a relation is called an *equivalence relation*.

Example 80. In organising a personal telephone list, it is convenient to organise the set S of people's names into 26 sets corresponding to the first letter in the name. In other words, we are making a section of the telephone list for each letter of the alphabet.

Example 81. Given a genealogy database, we would expect to see many queries about membership in families. We might assume that persons a and b are in the same family if they have the same last name, or there might be some other way to define what a family is (but it needs to have the property that *every* person belongs to *exactly one* family). One common query might be 'Is a in the same family as b?'. Once the *InSameFamilyAs* relation is defined, it provides all the information needed to organise all people into families.

These examples suggest that a relation we are using in this way needs to be reflexive (everything belongs in the same category as itself), symmetric (if a is in the same category as b, then obviously b must be in the same as a), and transitive (if a and b are in the same category, and so are b and c, then a and c must also be in the same category). These observations lead to the formal definition of an equivalence relation:

Definition 50. A binary relation R over a set A is an *equivalence relation* if it is reflexive, symmetric and transitive.

Example 82. Suppose that everyone in the database lives in a real location somewhere in the world. We can represent the world as a map, and then partition the map into small areas by using a *LivesIn SameLocationAs* relation.

The equivalence classes of a non-empty equivalence relation can be thought of as a *partition* of the set into disjoint subsets. Now we define this term formally:

Definition 51. A *partition* P of a non-empty set A is a set of non-empty subsets of A such that

- For each subset S_1 and S_2 of P, either $S_1 = S_2$ or $S_1 \cap S_2 = \emptyset$;

- $A = \bigcup_{S \in P} S$.

Example 83. For example, let's consider the set of people's last names and the relation `HasNameStartingWithSameLetterAs`. This relation divides the set into 26 subsets. There can be no overlaps between the subsets, and the set of names is the union of the subsets.

Example 84. Computer keyboards can generate several characters from most of the keys, depending on whether the Control, Shift or Meta keys are already down. We could define a relation *IsOnSameKeyAs*, which would partition the set of ASCII characters into equivalence classes, one for each key.

A good example of an equivalence relation, which is frequently used in computing applications, comes from the mathematical *modulus* (mod) operation on integers. The expression $e \bmod k$ gives the remainder produced when dividing e by k; and the value of $e \bmod k$ is a number between 0 and $k - 1$. Now, every number x which is a multiple of k will have the property that $x \bmod k = 0$, and we can build a set of all these numbers. Similarly, there is a set of numbers that have the same remainder when divided by 1, by 2, and so on when divided by k. To do this, we define the *congruence* relation as follows:

Definition 52. Let k be a positive integer, and let a and b be integers. If there is an integer n such that

$$(a - b) = n \times k,$$

Figure 8.25: A Congruence Equivalence Relation

then a is congruent to b (modulo k). The mathematical notation for this statement is

$$a \equiv b(\bmod k).$$

It is necessary to ensure that k is a *positive* integer—positive means greater than 0—in order to avoid dividing by 0. We can define a relation, called the congruence relation, for all k:

Definition 53. The congruence relation C_k is defined for all natural k such that $k > 0$, as follows: aC_kb if and only if $a \equiv b(\bmod k)$.

The congruence relation is useful because it is an equivalence relation:

Theorem 65. For all natural $k > 0$, the congruence relation C_k is an equivalence relation.

Example 85. Consider partitioning the integers by congruence (C_3) (see Figure 8.25). This gives rise to three sets: all the integers that are of the form $n \times 3$ (this is just the set of multiples of 3), the integers of the form $n \times 3 + 1$, and the integers of the form $n \times 3 + 2$.

The following functions in the software tools file create the smallest possible equivalence relation from a digraph, and determine whether a given relation is an equivalence relation. They do this by taking a digraph and calculating its transitive symmetric reflexive closure.

```
equivalenceRelation ::
  (Eq a, Show a) => Digraph a -> Digraph a
isEquivalenceRelation ::
  (Eq a, Show a) => Digraph a -> Bool
```

Exercise 42. Evaluate the following expressions using the computer:

```
equivalenceRelation ([1,2],[(1,1),(2,2),(1,2),(2,1)])
equivalenceRelation ([1,2,3],[(1,1),(2,2)])
```

```
isEquivalenceRelation ([1,2],[(1,1),(2,2),(1,2),(2,1)])
isEquivalenceRelation ([1],[])
```

Exercise 43. Does the topological sort require that the graph's relation is a partial order?

Exercise 44. Can the graph given to a topological sort have cycles?

8.10   Suggestions for Further Reading

Extensive discussions of relations can be found in *Discrete Mathematics in Computer Science*, by Stanat and McAllister [27] and *Discrete Mathematics*, by Ross and Wright [24].

Relations are a basic tool used for a wide variety of applications. Two good examples are relational data bases [6] and circuit design [18].

8.11   Review Exercises

Exercise 45. Which of the following relations is an equivalence relation?

(a) *InTheSameRoomAs*

(b) *IsARelativeOf*

(c) *IsBiggerThan*

(d) The equality relation

Exercise 46. Given a non-empty antisymmetric relation, does its transitive closure ever contain symmetric arcs?

Exercise 47. What relation is both a quasi order and an equivalence relation?

Exercise 48. Write a function that takes a relation and returns True if that relation has a power that is the given relation.

Exercise 49. A quasi order is transitive and irreflexive. Can it have any symmetric loops in it?

Exercise 50. Given an antisymmetric irreflexive relation, could its transitive closure contain reflexive arcs?

Exercise 51. Write a function that takes a relation and returns True if all of its powers have fewer arcs than it does.

Exercise 52. Write a function that takes a relation and returns True if the relation is smaller than its symmetric closure.

Exercise 53. Given the partial order

$$\{(A, B), (B, C), (A, D)\},$$

which of the following is not a topological sort?

```
[D,C,B,A]
[C,B,D,A]
[D,C,A,B]
```

Exercise 54. Is a reflexive and symmetric relation ever antisymmetric as well?

Exercise 55. Write a function that takes a relation containing a single path and calculates the transitive arcs added by its transitive closure. The length of a path is the number of arcs in it.

Exercise 56. Given a relation containing only a single path of length n, how many arcs can be added by its symmetric transitive closure?

Exercise 57. Given a relation containing only a cycle of length n containing all of the nodes in the domain, which power will be reflexive?

Exercise 58. Can we write a function that determines whether the equality relation over the positive integers is reflexive?

Exercise 59. Why can't partial orders have cycles of length greater than 1?

Exercise 60. Is the last power of a relation always the empty set?

Exercise 61. The following list comprehension gives the arcs of a poset diagram. What kind of order relation does the diagram represent?

```
[(a,a+1) | a <- [1..]]
```

Exercise 62. Is the composition of a relation containing only a single cycle with its converse the equality relation?

Exercise 63. Give examples of partial orders in which the set of greatest elements is the same as the set of weakest elements.

Chapter 9

Functions

A function is an abstract model of computation: you give it some input, and it produces a result. The essential aspect is that the result is completely determined by the input: if you repeatedly apply the same function to the same argument, you will always obtain the same result. Examples of functions include:

- An inquiry to a telephone directory service: you supply a person's name, and the service provides the corresponding telephone number;

- The mathematical sin function: you give it an angle, and the function returns the sin of that angle;

- An addition circuit in a computer's processor; you give it a pair of binary numbers, and it returns the binary representation of their sum.

In contrast, a weather prediction service would *not* be modelled as a function, since the answer to 'What will the weather be tomorrow?' changes from day to day.

Functions are an important tool, both in mathematics and in computing, because they provide a mechanism for abstraction. A function is a 'black box': to use it, you need to know the interface, but not the internal details about how the function is defined.

There are many ways to formalise the function concept. This chapter looks at the ones that are most important for computer science. We start with one of the most common mathematical approaches, which treats a function as a special kind of relation, and then we will consider a more algorithmic way of defining functions. It is good to remember, however, that the concept of 'function' is abstract, and there are many different formalisms that can be used to define it.

9.1   The Graph of a Function

In this section we examine one of the most common mathematical techniques for defining functions. This approach uses a set of ordered pairs, and it brings out a close connection between functions and relations.

A function specifies, for any particular input value x, the value y of the result. This can be represented as an ordered pair (x, y), and the entire function is represented as a set of ordered pairs. This representation of a function is called a *function graph*, and it is similar to the digraph used to represent relations.

A function is a relation, but some additional properties are required. A relation digraph is a set of ordered pairs, with no restrictions: thus, a relation R might contain two ordered pairs (x, y_1) and (x, y_2), and we would write $x\, R\, y_1$ and also $x\, R\, y_2$. This would not be acceptable for a function, though, since it would mean that given the argument x there are two possible results y_1 and y_2. A function must return a unique result for any argument. Accordingly, we define a function to be a relation with an extra requirement that only one result may be specified for each argument:

Definition 54. Let A and B be sets. A function f with type $A \to B$ is a relation with domain A and codomain B, such that

$$\forall x \in A.\ \forall y_1 \in B.\ \forall y_2 \in B.\quad ((x, y_1) \in f \land (x, y_2) \in f) \to y_1 = y_2.$$

A is called the *argument type* and B is called the *result type* of the function.

The definition says that a function is a set of ordered pairs, just like a relation. The set of ordered pairs is called the *graph* of the function. If an ordered pair (x, y) is a member of the function, then $x \in A$ and $y \in B$. Furthermore, the definition states formally that if the result of applying a function to an argument x could be y_1 but it could alternatively be y_2, then it must be that $y_1 = y_2$. This is just a way of saying that there is a unique result corresponding to each argument.

Example 86. The set $\{(1, 4), (1, 5)\}$ cannot be the graph of a function, because it contains two pairs with the same first element but different second elements: $(1, 4)$ and $(1, 5)$ (Figure 9.1). A function must return just one result for any argument; it can't choose among several alternatives.

Example 87. The set of ordered pairs $\{(1, 2),\ (2, 2),\ (3, 4)\}$ is the graph of a function. It doesn't matter that several arguments, 1 and 2, produce the same result 2.

An expression denoting the result produced by a function when presented with an input x is called a *function application*. For example, $\sin(2 \times \pi)$ is an application of the sin function to the argument $2 \times \pi$. The following definition specifies the syntax, type and value of a function application:

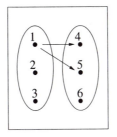

Figure 9.1: A Relation that is Not a Function

Definition 55. An *application* of the function f to the argument x, provided that $f :: A \rightarrow B$ and $x :: A$, is written as either $f\,x$ or as $f(x)$, and its value is y if the ordered pair (x, y) is in the graph of f; otherwise the application of f to x is undefined:

$$f\,x\ =\ y \quad \leftrightarrow \quad (x, y) \in f$$

A type can be thought of as the set of possible values that a variable might have. Thus the statement '$x :: A$', which is pronounced 'x has type A', is equivalent to $x \in A$. The type of the function is written as $A \rightarrow B$, which suggests that the function takes an argument of type A and transforms it to a result of type B. This notation is a clue to an important intuition: the function is a black box (you can think of it as a machine) that turns arguments into results.

If $x \in A$ and there is a pair $(x, y) \in f$, then we say that '$f\,x$ is defined to be y'. However, if $x \in A$ but there is no pair $(x, y) \in f$, we say '$f\,x$ is undefined'. A shorthand mathematical notation for saying '$f\,x$ is undefined' is '$f\,x = \bot$', where the symbol $\bot$ denotes an undefined value.

It often turns out that some of the elements of A and B don't actually appear in any of the ordered pairs belonging to a function graph. The subset of A consisting of arguments for which the function is actually defined is called the *domain* of the function. Similarly, the subset of B consisting just of the results that actually can be returned by the function is called the *image*. These sets are defined formally as follows:

Definition 56. The domain and the image of a function f are defined as:

$$domain\ f\ =\ \{x \mid \exists y.\ (x, y) \in f\}$$
$$image\ f\ =\ \{y \mid \exists x.\ (x, y) \in f\}$$

This definition says that the domain of a function is the set of all x such that (x, y) appears in its graph, while its image is the set of all y such that (x, y) appears. Thus a function must be defined for every element of its domain, and it must be able to produce every element of its image (given the right

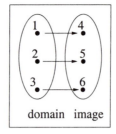

Figure 9.2: Domain and Image of a Function

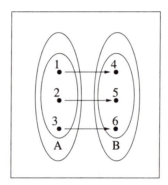

Figure 9.3: Function Type

argument). However, the argument type and the result type may contain extra elements that do not appear in the function graph (Figure 9.3).

Unfortunately, the terminology for functions is not entirely standard. Many authors use 'codomain' to refer to a function's argument type, but others define the codomain of a function differently. Many authors define the *range* of a function to be its image; others define it to be the result type. Whenever you are reading a document that uses any of these terms, you need to check the definitions given *in that document*.

Example 88. The set $\{(1,4),(2,5),(3,6)\}$ is the graph of a function (Figure 9.2). The domain is $\{1,2,3\}$ and the image is $\{4,5,6\}$. The function can have any type of the form $A \to B$, provided that $\{1,2,3\} \subseteq A$ and $\{4,5,6\} \subseteq B$.

Example 89. Let function $f :: Integer \to Integer$ be defined as

$$f = \{(0,1),(1,2),(2,4),(3,8)\}.$$

The argument type of f is $Integer = \{\dots,-2,-1,0,1,2,\dots\}$, and its domain

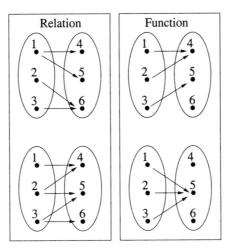

Figure 9.4: Two Relations and Two Functions

is $\{0, 1, 2, 3\}$. The result type is *Integer* $= \{\ldots, -2, -1, 0, 1, 2, \ldots\}$, and the image is $\{1, 2, 4, 8\}$.

Example 90. Figure 9.4 shows the graph diagrams for two relations and two functions.

9.2 Functions in Programming

The 'function as a graph' idea used to model a function mathematically is not exactly the same as a function written in a programming language, although both are realisations of the same idea. The difference is that a set of ordered pairs specifies *only* what result should be produced for each input; there is no concept of an algorithm that can be used to obtain the result. The function graph approach 'pulls the result out of a hat'. In contrast, a function in a programming language is represented solely by the algorithm, and the only way to determine the value of $f\ x$ is to execute the algorithm on input x. A programming language function is a method for computing results; a mathematical function is a set of answers.

Besides providing a method for obtaining the result, a programming language function has a behaviour: it consumes memory and time in order to compute the result of an application. For example, we might write two sorting functions, one that takes very little time to run on a given test sample and one that takes a long time. We would regard them as different algorithms, and would focus on that difference as being important. The graph model of a function lacks any notion of speed, and would make no distinction between the two algorithms as long as they always produce the same results.

There are several important classes of functions defined by algorithms, which we will examine in the next few sections. The essential questions we are interested in are the termination and execution speed of the function.

9.2.1 Inductively Defined Functions

As its name suggests, inductively defined functions use a computation structure that is similar to induction, which can be used to prove properties of these functions.

Definition 57. A function defined in the following form, where h is a non-recursive function, is *inductively defined*:

$$f\,0 \; = \; k$$
$$f\,n \; = \; h(f\,(n-1))$$

When the argument is 0, the function returns a constant k, and when the argument n is positive, it calls itself recursively on a smaller argument $n-1$; the function can then use h to perform further calculations with the result of the recursive call.

As long as h always returns a result, an inductively defined function will always produce a result when applied to any nonnegative argument.

Example 91. The function defined below the sum $\sum_{i=0}^{n} i$, counting backwards from n down to 0. It is written in the form required for inductively defined functions, letting $k = 0$ and $h\,x = n + x$.

```
f 0 = 0
f n = n + f (n-1)
```

Example 92. The add function, defined below, is inductively defined over the second argument.

```
add :: Int -> Int -> Int
add n 0 = 0
add n k = n + add n (k - 1)
```

Example 93. The '91' function is recursive, but is not inductively defined: it calls itself recursively on a *larger* argument, and it performs yet another recursion on the result (f91 (x+11)) of the first recursion. When applied to n, this function returns 91 if $0 \le n \le 100$, and otherwise it returns $n - 10$.

```
f91 :: Int -> Int
f91 x = if x > 100
           then x - 10
           else f91 (f91 (x + 11))
```

9.2.2 Primitive Recursion

Computability theory is the branch of computer science that studies the properties of functions viewed as algorithms. One of the most important classes of algorithm is the set of *primitive recursive functions*, which are defined with a more flexible pattern than inductively defined functions. Primitive recursive functions are essentially equivalent to algorithms that can be expressed with looping structures, such as **for** loops, that are guaranteed to terminate.

Definition 58. A function f is *primitive recursive* if its definition has the following form, where g and h are primitive recursive functions.

$$f\ 0\ x\ =\ g\ x$$
$$f\ (k+1)\ x\ =\ h\ (f\ k\ x)\ k\ x$$

This definition specifies the standard form for a primitive recursive function. Any function which can be transformed into this form is primitive recursive, even if its definition doesn't obviously match the definition.

Example 94. The sqr function, which takes a natural number x and returns x^2, is primitive recursive, as shown by the following definition. This function satisfies the requirements vacuously, since it does not actually use recursion. All 'basic functions' that do not require recursion can be handled the same way, so they are all primitive recursive.

```
sqr x = f 0 x
   where f 0 x = g x
         g x = x*x
```

Example 95. The factorial function can be written in the standard primitive recursive form:

```
factorial k = f k undefined
   where f 0 x = 1
         f (k+1) x = (k+1) * (f k x)
```

The function f performs a recursion over k, starting from the argument to factorial and counting down to 0. Since factorial can be calculated simply by multiplying together all the numbers in this countdown, the x argument is not actually required, and f ignores the value of x. The definition of factorial could pick any arbitrary value for x, and undefined $= \perp$ is as good as any. Since x is not used, factorial doesn't make full use of the power of primitive recursion. The following calculation shows how the application factorial 4 can be reduced to 24 (notice that $\perp$ is repeatedly copied but never used):

$$
\begin{aligned}
& \quad \textit{factorial } 4 \\
&= \ f\ 4\ \perp \\
&= \ 4 \times f\ 3\ \perp \\
&= \ 4 \times (3 \times f\ 2\ \perp) \\
&= \ 4 \times (3 \times (2 \times f\ 1\ \perp)) \\
&= \ 4 \times (3 \times (2 \times (1 \times f\ 0\ \perp))) \\
&= \ 4 \times (3 \times (2 \times (1 \times 1))) \\
&= \ 4 \times (3 \times (2 \times 1)) \\
&= \ 4 \times (3 \times 2) \\
&= \ 4 \times 6 \\
&= \ 24
\end{aligned}
$$

Example 96. The following function is not primitive recursive, because if the argument is odd (except for 1) it calls itself recursively with the same arguments. However, if the argument is a power of 2 then the recursion will terminate with f 1, and the function will return the logarithm (base 2) of its argument.

```
f 0 = 0
f 1 = 0
f x =
  if even x
  then 1 + f (x 'div' 2)
  else f x
```

9.2.3 Computational Complexity

The *computational complexity* of a function is a measure of how costly it is to evaluate. The memory consumption and the time required are common measures of the cost of a function.

Recursion can create some very expensive computations. A famous example is Ackermann's function:

Definition 59. Ackermann's function is:

```
ack 0 y = y+1
ack x 0 = ack (x-1) 1
ack x y = ack (x-1) (ack x (y-1))
```

The ack function is easy to evaluate for small arguments, but the time it takes grows extremely quickly as x and y increase. Books on computability theory and algorithmic complexity show why this happens, but it is interesting to make a table for yourself of ack x y for small values of the arguments.

9.2.4 State

A function always returns the same result, given the same argument. This kind of repeatability is essential: if $\sqrt{4} = 2$ today, then $\sqrt{4} = 2$ also tomorrow. Some computations do not have this property. For example, many programming languages provide a 'function' that returns the current date and time of day, and the result returned from such a query will definitely be different tomorrow. The entire set of circumstances that can affect the result of a computation is called the *state*.

Example 97. The state of a computer system includes the current date and time of day, as well as the contents of the file system. Thus a 'function' that queries the date, or the amount of free space on disk, will not return the same result every time it is called. These are not true functions, although some programming languages use the keyword `function` erroneously to refer to them.

Example 98. As consumers, we expect to have to trade money for products. The interface between us and those that sell these products is a functional one. However, we also have to take into consideration things like depreciation over time, or wear and tear, or product expiration dates. These issues concern the state of the items for which we trade money.

Example 99. Some programming languages allow a 'function' to modify the value of a global variable. Even if such a 'function' always returns the same result for each argument value, its behaviour is not in principle describable by a function graph. A 'function' in a program that modifies the global state is not a mathematical function.

Notwithstanding these examples, it is possible to describe computations with state using pure mathematical functions. The idea is to include the state of the system as an extra argument to the function. For example, suppose we need a function f that takes an integer and returns an integer, but the result might also depend on the state of the computer system (perhaps the time of day, or the contents of the file system). We can handle this by defining a new type `State` that represents all the relevant aspects of the system state, and then providing the current state to `f`:

```
f :: State -> Int -> Int
```

Now f can return a result that depends on the time of day, even though it is a mathematical function. Given the same system state and the same argument, it will always return the same result.

Programs that need to manipulate the state can be written as pure functions, with the state made explicit and passed as an argument to each function that uses it. When a program needs to use the state frequently, however, it becomes awkward to use explicit `State` arguments; this clutters up the program and errors in keeping track of the state can be hard to find.

Imperative languages solve this problem by making the state implicit, and allowing *side effects* to modify the state. This is a simple way to allow algorithms to use the system state. The cost of this approach is that reasoning about the program is more difficult. In mathematics, if you have an equation of the form $x = y$, you can replace x by y, or vice versa. This is called *substituting equals for equals* or *equational reasoning*. Unfortunately, equational reasoning doesn't work in general in imperative languages. If a function f depends on the system state, it is not even true that f x = f x. It is still possible to reason formally about imperative programs, for example using the *weakest precondition method*, but this is more complex than equational reasoning.

Haskell takes a different approach: it provides a mechanism allowing you to define operations that use the state implicitly. The mechanism is called a monad, and it is used with do expressions in Haskell. The technique is explained in [30].

9.3   Higher Order Functions

A distinction is often made between data (numbers, characters, etc.) and algorithms (code to be executed on a computer). For example, 23 and [1,2,3] are data values, while the `length` function is code to be executed. Many programming languages treat functions as code, and disallow their use as data. This means that the arguments and result of a function application must be data; functions themselves cannot be used as arguments or results.

In both mathematics and functional programming languages, this restriction is removed: computer code—in the form of functions—can be used as ordinary data values. This means, for example, that you can store functions in data structures. You can also pass a function like `length` to some other function, which might use it; you can also write a function that does some computation, and then produces a brand new function which it returns. Functions like this are called *higher order functions*.

Definition 60. A *first order function* has ordinary (non-function) arguments and result. A *higher order function* is one that either takes a function as an argument, or returns a function as its value, or both.

Example 100. The Haskell function `map` is a higher order function, because its first argument is a function which `map` will apply to each element of the second argument, which is a list. The type of `map` is:

```
map :: (a->b) -> [a] -> [b]
```

This type reveals that the function is higher order, since the first argument has a type that contains an arrow, indicating that this argument is a function.

Example 101. The `length` function is not higher order. As its type makes plain, the argument is a list type and the result is an integer:

```
length :: [a] -> Int
```

We will now look in detail at the various kinds of higher order functions: functions that take other functions as arguments and functions that return functions as results. We will also compare two methods for allowing a function to take several arguments: building a tuple so that all the arguments are packaged in a data structure, and using higher order functions to take the arguments one at a time.

9.3.1 Functions that Take Functions as Arguments

Any function that takes another function as an argument is higher order. This kind of higher order function will have a type something like the following:

$$f :: (\cdots \rightarrow \cdots) \rightarrow \cdots$$

We have already seen many examples of such functions in Haskell; map and foldr are typical. Generally, this variety of higher order function will also take a data argument, and it will apply its function argument to its data argument in a special way.

Example 102. As we have already seen, the map function takes a data structure (which must be a list of data values) and applies its function argument to each element of the list.

```
map :: (a->b) -> [a] -> [b]
map f [] = []
map f (x:xs) = f x : map f xs
```

Example 103. The following function performs an operation twice on the data argument x. The operation to be performed is specified by the first argument f, which is a function.

```
twice :: (a -> a) -> a -> a
twice f x = f (f x)
```

Notice that the type of f is more restricted in twice than it is in map. The reason for this is that nothing in map constrains either the argument type or the result type of f, so the function can have the general function type a->b. In twice, however, the result returned by f is used as the argument to another application of f. This means that f must have the less general function type a->a.

Higher order functions provide a flexible and powerful approach to user-defined control structures. A *control structure* is a programming language construct that specifies a sequence of computations. Examples of control structures in imperative programming languages include **for** loops, **while** loops, the **if** statement, and the like.

It is often possible to define a higher order function which implements a control structure. For example, let ys be a list of length n. Then the Haskell equation

```
ys = map f xs
```

is similar to the following **for** loop (written in a common imperative style):

```
for i = 1 to n
  y[i] := f (x[i]);
```

Thus map describes an iteration that computes a list of values, where the ith element of the result is computed from the ith element of the argument by applying the function f. Similarly,

```
y = foldl f a xs
```

is similar to the following imperative loop:

```
y := a;
for i := 1 to n do
  y := f y x[i]
```

9.3.2   Functions that Return Functions

Any function that returns another function as its result is higher order, and its type will have the following form:

$$ f :: \cdots \to (\cdots \to \cdots) $$

To understand this kind of higher order function, it is helpful to study the function graph in detail. First, we define some first order functions to be used in the examples:

```
ident, double, triple, quadruple :: Int -> Int

ident 1 = 1
ident 2 = 2
ident 3 = 3

double 1 = 2
double 2 = 4
double 3 = 6

triple 1 = 3
triple 2 = 6
triple 3 = 9
```

```
quadruple 1 = 4
quadruple 2 = 8
quadruple 3 = 12
```

These simple functions perform multiplication on small arguments. For example, `double` takes a number x (which must be 1, 2 or 3) and it returns $2 \times x$. The graph of a first order function is a straightforward set of ordered pairs; the functions just defined have the following graphs:

$$
\begin{aligned}
\texttt{ident} &= \{(1,1),(2,2),(3,3)\} \\
\texttt{double} &= \{(1,2),(2,4),(3,6)\} \\
\texttt{triple} &= \{(1,3),(2,6),(3,9)\} \\
\texttt{quadruple} &= \{(1,4),(2,8),(3,12)\}
\end{aligned}
$$

Now we define a higher order function, `multby`, which takes one argument of type `Int` and returns a function with type `Int->Int`:

```
multby :: Int -> (Int->Int)
multby 1 = ident
multby 2 = double
multby 3 = triple
multby 4 = quadruple
```

This function simply looks at its first argument x, an integer which must be in $\{1,2,3,4\}$, and it returns another function. Now consider the value of `multby 3 2`. This is syntactically equivalent to `(multby 3) 2`, and we can evaluate the expression using the function definitions. Notice that `multby 3` returns a function which multiplies things by 3, so `multby 3 2` evaluates to 6:

```
multby 3 2
   = (multby 3) 2      syntax rule of Haskell
   = triple 2          definition of multby (third equation)
   = 6                 definition of triple (second equation)
```

If a function returns another function as its result, then its graph will be a set of ordered pairs (x, y) where x is the argument to the function and y is another function graph. Figure 9.5 shows the graph of `multby`:

$$
\begin{aligned}
\texttt{multby} = \{&(1,\ \{(1,1),(2,2),(3,3)\}) \\
&(2,\ \{(1,2),(2,4),(3,6)\}) \\
&(3,\ \{(1,3),(2,6),(3,9)\}) \\
&(4,\ \{(1,4),(2,8),(3,12)\})\}
\end{aligned}
$$

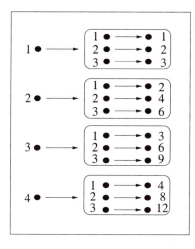

Figure 9.5: Graph of the `multby` Higher Order Function

9.3.3   Multiple Arguments as Tuples

Technically, a function (in either mathematics or Haskell) takes exactly one argument and returns exactly one result. There are two ways to get around this restriction. One method is to package multiple arguments (or multiple results) in a tuple. Suppose that a function needs two data values, x and y. The caller of the function can build a pair (x, y) containing these values, and that pair is now a single object which can be passed to the function.

Example 104. The following function takes two numbers and adds them together:

```
add :: (Integer,Integer) -> Integer
add (x,y) = x+y
```

The function is called using an application that builds a suitable tuple. Thus `f (3,4)` applies `add` to the pair `(3,4)`; when `(x,y)` is matched with `(3,4)`, the effect is to define `x` to be 3 and `y` to be 4 in the body of the function. The graph of the function is an infinite set, since `Integer` is a type with an infinite number of values. The graph has the following form:

```
add =
  { ··· ,
    ··· ,  ((0,−2),−2),  ((0,−1),−1),  ((0,0),0),  ((0,1),1),  ...
    ··· ,  ((1,−2),−1),  ((1,−1),0),   ((1,0),1),  ((1,1),2),  ...
    ··· ,  ((2,−2),0),   ((2,−1),−1),  (2,0),2),   ((2,1),3),  ...
    ···
  }
```

9.3.4 Multiple Results as a Tuple

A function must return exactly one result, but sometimes in practice we want one to return several pieces of information. In such cases, the multiple pieces can be packaged into a tuple, and the function can return that as a single result. This technique is analogous to passing several arguments as a tuple.

Example 105. The following Haskell function takes an integer x, and returns two results $x - 1$ and $x + 1$ packaged in a pair (i.e. a 2-tuple):

```
addsub1 :: Integer -> (Integer,Integer)
addsub1 x = (x-1, x+1)
```

The function graph is:

```
addsub1 =
  {  ... ,
```
$$(-2, (-3, -1)), \quad (-1, (-2, 0)), \quad (0, (-1, 1)), \quad (1, (0, 2)),$$
$$(2, (1, 3)), \qquad\quad (3, (2, 4)), \qquad (4, (3, 5)), \qquad (5, (4, 6)),$$
```
  ...
  }
```

9.3.5 Multiple Arguments with Higher Order Functions

Higher order functions provide another method for passing several arguments to a function. Suppose that a function needs to receive two arguments, $x :: a$ and $y :: b$, and it will return a result of type c. The idea is to define the function with type

$$f :: a \rightarrow (b \rightarrow c).$$

Thus f takes only one argument, which has type a, and it returns a function with type $b \rightarrow c$. The result function is ready to be applied to the second argument, of type b, whereupon it will return the result with type c. This method is called *Currying*, in honour of the logician Haskell B. Curry (for whom the programming language Haskell is also named).

The graph of f is a set of ordered pairs; the first element of each pair is a data value of type a, while the second element is the function graph for the result function. That function graph contains, in effect, the information that f obtained from the first argument.

Example 106. The following function, mult, is similar to add (see example 104); apart from using * rather than +, the only difference is that mult is higher order, and takes its arguments one at a time:

```
mult :: Integer -> (Integer->Integer)
mult x y = x*y
```

The graph of mult is a set of ordered pairs of the form (k, f_k), where k is the value of the first argument to mult, and f_k is the graph of a function that takes a number and multiplies it by k:

$$
\begin{aligned}
\texttt{mult} = {} & \\
\{ & \cdots, \\
(-1, & \quad \{\ldots, \quad (-1,1), \quad (0,0), \quad (1,-1), \quad (2,-2), \quad \ldots\}), \\
(0, & \quad \{\ldots, \quad (-1,0), \quad (0,0), \quad (1,0), \quad (2,0), \quad \ldots\}), \\
(1, & \quad \{\ldots, \quad (-1,-1), \quad (0,0), \quad (1,1), \quad (2,2), \quad \ldots\}), \\
(2, & \quad \{\ldots, \quad (-1,-2), \quad (0,0), \quad (1,2), \quad (2,4), \quad \ldots\}), \\
(3, & \quad \{\ldots, \quad (-1,-3), \quad (0,0), \quad (1,3), \quad (2,6), \quad \ldots\}), \\
& \cdots \\
\}
\end{aligned}
$$

9.4 Total and Partial Functions

Recall that the domain of a function $f :: A \to B$ is a subset of A consisting of all the elements of A for which f is defined. There are two sets that can be used to describe the possible arguments of f. The argument type A is generally thought of as a constraint: if you apply f to x, then it is required that $x \in A$ (alternatively, $x :: A$). If this constraint is violated then the application $f\ x$ is meaningless. The domain of f is the set of arguments for which f will produce a result. Naturally, domain f must be a subset of A. However, there is an important distinction between functions where the domain is the same as the argument type, and functions where the domain is a proper subset of the argument type.

Definition 61. Let $f :: A \to B$ be a function. If domain $f = A$ then f is a *total* function. If domain $f \subset A$ then f is a *partial* function.

If f is a partial function, and $x \in$ domain f, then we say that $f\ x$ is defined. There are several standard ways to describe an application $f\ y$ where $y :: A$ but $y \notin$ domain f. It is common, especially in mathematics, to write '$f\ y$ is undefined'. Another approach, frequently used in theoretical computer science, is to introduce a special symbol $\perp$ (pronounced bottom) which stands for an undefined value. This allows us to write $f\ y = \perp$.

Example 107. The following function has argument type Integer, but its domain is $\{1, 2, 3\}$:

```
f :: Integer -> Char
f 1 = 'a'
f 2 = 'b'
f 3 = 'c'
```

The expression f 1 is defined, and its value is the character 'a'. The expression f 4 is undefined, and it has no value. Another way to say this is that f 1 = 'a' and f 4 = $\perp$. The graph of f is $\{(1, \text{'a'}), (2, \text{'b'}), (3, \text{'c'})\}$.

Partial functions are useful when describing the behaviour of programs. Generally, the type of a function can be used for compile-time analysis by the compiler, but the domain may be difficult or impossible to work out from the function's definition.

Example 108. The function `sqrt ::  Float->Float` takes the square root of its argument. The argument type is Float, and the compiler uses that constraint to detect errors in the program. Thus if a program contains an application like `sqrt "cat"` the compiler will produce an error message, since a character string is not an element of the type `Float`. However, the compiler cannot determine whether a numeric argument to `sqrt` will be positive or negative.

Example 109. The following function can be applied to any integer, but it is defined only for 1:

```
justone :: Int -> Int
justone 1 = 3
```

If this function is applied to anything that isn't an integer, the compiler will produce a type error message, and the program cannot be executed. If it is applied to 2, no error is detected at compile time but a runtime error message will be produced, for example:

```
> justone 2
```

```
Program error: {justone 2}
```

Many programming languages have the property that some type errors are not detectable by the compiler, and the application of a function to an argument of the wrong type is likely to crash the program. The Haskell type system is designed carefully so that the compiler guarantees that all type errors will be detected at compile time; it is impossible for the program to crash at runtime due to a type error. Unix programmers using C are accustomed to running a program and getting the message `segmentation fault`; this is caused by a type error (for example, if an integer value is used as an address). Such errors are rare in Haskell; even though the compiler knows nothing about the domains of functions, it is able to catch most errors just by checking their types.

If a function is applied to a value which has the right type, but which is not in the function's domain, then a runtime error occurs. Sometimes the program, or the system, is able to detect this and produce an error message. For example, if the `sqrt` function is applied to -2, an error message will be produced explaining what happened.

Some runtime error messages are generated automatically, but Haskell also allows you to implement such error messages yourself. The function `error` takes a string argument, which is an error message; if an application of `error` is evaluated, the string is printed and the program execution is terminated.

Example 110. The argument type of the following function is `Integer`, but its domain is the set of non-negative integers. Suppose that it would be a runtime error to evaluate an application of `f` to a negative number, but suppose also that we would like an informative error message if this happens. The `error` function can be used to terminate the execution with a tailor-made message. A common technique, illustrated here, is to construct the error message string, including pertinent information about the argument. In this case, we simply convert x to a string, with `show x`, and include that in the message.

```
f :: Integer -> Integer
f x =
  if x >= 0
    then 10*x
    else error ("f was applied to a negative number "
                ++ show x ++ ".  Don't do it again!")
```

Here are the results of evaluating two applications of `f`. The argument is in the domain of `f` in the first application, but not in the second.

```
> f (2+2)
40
> f (2-5)

Program error: f was applied to a negative
  number -3.  Don't do it again!
```

Unfortunately, it is not possible to detect all errors at runtime, and when an undetectable error occurs it is impossible to print a useful error message. Sometimes a recursive function goes into an infinite loop, and there is simply no output at all.

Example 111. The `infinite_loop` function takes an argument and calls itself with the same argument. Since it makes no progress toward a termination condition, any application of this function will run forever. The user must interrupt the execution, typically by typing CTRL-C or by clicking on Stop.

```
infinite_loop :: a -> a
infinite_loop x = infinite_loop x
```

An application of a total function will always terminate and produce a result. There are three possible outcomes from evaluating an application of a partial function to an argument: the application could terminate with a result; it could produce a runtime error message; or it could go into an infinite loop. For example, the following function will terminate if its argument is even, but it will go into an infinite loop if the argument is odd:

```
haltIfEven x =
  if even x
```

```
    then x
    else infinite_loop x
```

If we try applying `haltIfEven` to some arguments, we might quickly discover that `haltIfEven 6` ⇒ 6. However, when we attempt to evaluate the expression `haltIfEven 7` there will not be any output at all; the computation will just go on and on. When a computer is running a program but not producing any output, it would be useful to know whether it just needs more time, or whether it is stuck in an infinite loop. For this particular example it is obvious when the function will terminate and when it will loop forever, but what about more complicated functions where this is not obvious?

It would be extremely useful to have a function called `wouldHalt` that takes an *arbitrary* function `f` and an argument value `x`, and which returns True if and only if `f` would halt if we actually evaluate `f x`:

```
    wouldHalt :: (Integer->Integer) -> Integer -> Bool
```

This is called the *Halting Problem*.

Obviously we cannot implement `wouldHalt` by actually evaluating `f`. Suppose we write it something like this:

```
    wouldHalt :: (Integer->Integer) -> Integer -> Bool
    wouldHalt f x =
      if f x == f x
        then True
        else False
```

The problem is that if `f x` goes into an infinite loop, then `wouldHalt` will never get the opportunity to return **False**. It will always return **True** if the result should be **True**, but it will go into an infinite loop if the result should be **False**.

Since `wouldHalt` cannot actually evaluate `f x`, it must instead analyse the definition of `f`, somehow figure out how it works, and then decide whether `f x` would halt. We could easily define a simplified `wouldHalt` that can handle `haltIfEven` and similar functions. Could we extend it, with enough effort, so that our function solves the Halting Problem? Unfortunately this is impossible:

Theorem 66. There does not exist a function `wouldHalt` such that for all `f` and `x`,

$$\text{wouldHalt } f \ x = \begin{cases} \text{True,} & \text{if } f \ x \text{ terminates} \\ \text{False,} & \text{if } f \ x \text{ does not terminate} \end{cases}$$

Proof. Define the function `paradox` as follows:

```
    paradox :: Integer -> Integer
    paradox x =
      if wouldHalt paradox x
      then paradox x
      else 1
```

Now consider the expression `paradox x`. One of the following two cases must hold:

- Suppose `paradox x` halts and produces a result. Then

  wouldHalt paradox x ⇒ True;

  therefore the definition reduces to `paradox x = paradox x`, so `paradox x` does not halt. This is a contradiction; therefore it is impossible that `paradox x` halts.

- Suppose `paradox x` does not halt. Then

  wouldHalt paradox x ⇒ False.

  The definition then simplifies to `paradox x = 1`, and it halts.

To summarise, if `paradox x` halts then it does not halt, and if it does not halt then it halts! There is no possibility which avoids contradiction. Therefore the function `wouldHalt` does not exist.   □

The proof that the Halting Problem is unsolvable was discovered in the 1930s by Alan Turing. This is one of the earliest and most important results of computability theory. One of the commonest methods for proving that a function does *not* exist is to show how it could be used to solve the Halting Problem, which is unsolvable. This theorem also has major practical implications: it means that some software tools which would be very useful cannot actually be implemented.

A consequence of the unsolvability of the Halting Problem is that it is impossible to write a Haskell function that determines whether another Haskell function is total or partial. However, we can introduce data structures to represent the graphs of partial functions, with an explicit value that represents ⊥. The software tools file takes this approach, as described below.

The definition of a function requires that every element of the domain be mapped to some element of the result type, which can be the undefined value. This means that we need to use a new type to represent the result returned by a function, called `FunVals`. This type has two kinds of element. The first is `Undefined`, which means that the result value is ⊥. The other is called `Value`, and takes an argument which is the actual value returned by the function.

Now we can define a predicate that returns `True` if its argument is a partial function (in other words if some member of its result type is undefined), and `False` otherwise:

```
isPartialFunction
  :: (Eq a, Show a, Eq b, Show b)
  => Set a -> Set b -> Set (a,FunVals b) -> Bool
```

There is also a function isFun that takes a relation, and determines whether the relation is also a function:

```
isFun :: (Eq a, Show a, Eq b, Show b) =>
         Set a -> Set b -> Set (a,FunVals b) -> Bool
```

Exercise 1. Decide whether the following functions are partial or total, and then run the tests on the computer:

   (a)   isPartialFunction
```
            [1,2,3] [2,3]
            [(1,Value 2),(2,Value 3),(3,Undefined)]
```

   (b)   isPartialFunction
```
            [1,2] [2,3]
            [(1,Value 2),(2,Value 3)]
```

Exercise 2. Work out the following expressions, by hand and using the computer:

```
isFun [1,2,3] [1,2] [(1,Value 2),(2,Value 2)]
isFun [1,2,3] [1,2] [(1,Value 2),(2,Value 2),
                     (3,Value 2),(3,Value 1)]
isFun [1,2,3] [1,2] [(1,Value 2),
                     (2,Value 2),(3,Value 2)]
```

Exercise 3. What is the value of mystery x where mystery is defined as:

```
mystery :: Int -> Int
mystery x = if mystery x == 2 then 1 else 3
```

Exercise 4. What is the value of mystery2 x where mystery2 is defined as:

```
mystery2 :: Int -> Int
mystery2 x = if x == 20 then 2 + mystery2 x else 3
```

9.5 Function Composition

It is often possible to structure a computation as a sequence of function applications organised as a pipeline: the output from one function becomes the input to the next, and so on. In the simplest case there are just two functions in the pipeline: the input x goes into the first function, g, whose output goes into the second function f (Figure 9.6).

Definition 62. Let $g :: a \rightarrow b$ and $f :: b \rightarrow c$ be functions. Then the *composition* of f with g, written $f \circ g$, is a function such that:

$$(f \circ g) \quad :: \quad a \rightarrow c$$
$$(f \circ g)\, x \quad = \quad f\, (g\, x)$$

Figure 9.6: Functional Composition $(f \circ g)\ x\ =\ f\ (g\ x)$

When you think of composition as a pipeline, the input x goes into the first function g; this produces an intermediate result $g\ x$ which is the input to the second function f, and the final result is $f\ (g\ x)$. Notice, however, that in the notation $f \circ g$, the functions are written in backwards order. This may be unfortunate, but this is the standard definition of function composition, and you need to be familiar with it. Just remember that $f \circ g$ means *first* apply g, *then* f.

The $\circ$ symbol is an operator that takes two functions and produces a new function, just as $+$ is an operator that takes two numbers and produces a new number. The first argument to $\circ$ is $f :: b \rightarrow c$, and the second argument is $g :: a \rightarrow b$, and the entire composition takes an input $x :: a$ and returns a result $(f\ (g\ x)) :: c$. Therefore the $\circ$ operator has the following type:

$$(\circ)\ ::\ (b \rightarrow c) \rightarrow (a \rightarrow b) \rightarrow (a \rightarrow c)$$

Haskell has a built-in operator for function composition. Since $\circ$ is unfortunately not an ASCII character, the full stop character '.' is used instead to denote function composition.

Definition 63. The Haskell function composition operator is:

```
(.) :: (b -> c) -> (a -> b) -> (a -> c)
(f.g) x = f (g x)
```

Example 112. Suppose that you wish to increment the second elements of a list of pairs called lstpairs. You could write this as

```
map increment (map snd lstpairs)
```

but it is often clearer to write it as

```
map (increment.snd) lstpairs
```

This helps the reader to see that several operations are going to be applied in turn to each element of the list.

Composition is often used to define a processing pipeline, as shown in Figure 9.6, but we often need more than two functions in the pipeline. Figure 9.7 shows a typical example, where four functions are connected together using $\circ$. The following theorem states an extremely important property of the $\circ$ operator which makes it easier to define such pipelines:

Theorem 67. Functional composition ($\circ$) is associative. That is, for all functions $h :: a \to b, g :: b \to c, f :: c \to d$,

$$f \circ (g \circ h) \;=\; (f \circ g) \circ h,$$

and both the left and right hand sides of the equation have type $a \to d$.

Proof. We need to prove that the two functions $f \circ (g \circ h)$ and $(f \circ g) \circ h$ are extensionally equal. To do this we need to choose an arbitrary $x :: a$, and show that both functions give the same result when applied to x. That is, we need to prove that the following equation holds:

$$(f \circ (g \circ h)) \; x \;=\; ((f \circ g) \circ h) \; x.$$

This is done by straightforward equational reasoning, repeatedly using the definition of $\circ$:

$$
\begin{aligned}
&((f \circ g) \circ h) \; x \\
={}& (f \circ g) \; (h \; x) \\
={}& f \; (g \; (h \; x)) \\
={}& f \; ((g \circ h) \; x) \\
={}& (f \circ (g \circ h)) \; x
\end{aligned}
$$

$\square$

The significance of this theorem is that we can consider a complete pipeline as a single black box; there is no need to structure it two functions at a time. Similarly, you can omit the redundant parentheses in the mathematical notation. The following compositions are all identical:

$$
\begin{aligned}
&(f_1 \circ f_2) \circ (f_3 \circ f_4) \\
&f_1 \circ (f_2 \circ (f_3 \circ f_4)) \\
&((f_1 \circ f_2) \circ f_3) \circ f_4 \\
&f_1 \circ ((f_2 \circ f_3) \circ f_4) \\
&(f_1 \circ (f_2 \circ f_3)) \circ f_4
\end{aligned}
$$

The parentheses in all of these expressions have no effect on the meaning of the expression, and they make the notation harder to read, so it is customary to omit them and write simply

$$f_1 \circ f_2 \circ f_3 \circ f_4.$$

Notice, however, that you must put parentheses around an expression denoting a function when you apply it to an argument. For example,

$$(f_1 \circ f_2 \circ f_3 \circ f_4) \; x$$

is the correct way to apply the pipeline to the argument x. It would be incorrect to omit the outer parentheses, as in

$$f_1 \circ f_2 \circ f_3 \circ f_4 \; x,$$

Figure 9.7: A Pipeline Defined via Functional Composition $f_4 \circ f_3 \circ f_2 \circ f_1$

because function application takes precedence over all other operations, and the meaning would be equivalent to

$$f_1 \circ f_2 \circ f_3 \circ (f_4\ x),$$

which is not what was intended. In functional programming it is common to see a pipeline of functions connected with the $\circ$ operator (written as '.' in Haskell); if this expression is applied to an argument then there is just one pair of parentheses around the whole pipeline, but otherwise no parentheses are needed. For example, here are two Haskell equations, defining a function g and a data value y:

```
g = f1 . f2 . f3 . f4
y = (f1 . f2 . f3 . f4) x
```

The software tools contain an implementation of the following function; its two arguments are functions represented as graphs, and it returns the composition, also represented as a graph.

```
functionalComposition
  :: (Eq a, Show a, Eq b, Show b, Eq c, Show c)
  => Set (a,FunVals b) -> Set (b,FunVals c)
     -> Set (a,FunVals c)
```

Exercise 5. Work out the values of the following expressions, and then check your result by evaluating them with the computer:

```
map (increment.increment.increment) [1,2,3]
map ((+ 2).(* 2)) [1,2,3]
```

Exercise 6. Using the definitions below, work out the type and graph of f.g, and check using the computer.

```
f : Int -> String
f 1 = "cat"
f 2 = "dog"
f 3 = "mouse"

g : Char -> Int
g 'a' = 1
```

```
g 'b' = 2
g 'c' = 2
g 'd' = 3
```

Exercise 7. Functions are often composed with each other in order to form a pipeline that processes some data. What does the following expression do?

```
((map (+ 1)).(map snd)) xs
```

Exercise 8. Sometimes access to deeply nested constructor expressions is performed by function composition. What is the value of this expression?

```
(fst.snd.fst) ((1,(2,3)),4).
```

9.6 Properties of Functions

We have used four sets to characterise a function: the argument type, the domain, the result type, and the image. There are several useful properties of functions that concern these four sets, and we will examine them in this section.

9.6.1 Surjective Functions

A surjective function has an image which is the same as its result type (sometimes called the range). Thus the function can, given suitable input, produce any of the elements of the result type.

Definition 64. A function $f :: A \rightarrow B$ is *surjective* if

$$\forall b \in B. \ (\exists a \in A. \ f \ a = b).$$

Example 113. The function even :: Integer -> Bool has a result type with only two elements, True and False. Both of these values are in the function's image, as demonstrated by the applications even 2 = True and even 3 = False. Therefore every element of the result type is also an element of the image, and even is surjective.

Example 114. The times_two function takes an integer and doubles it:

```
times_two :: Int -> Int
times_two n = 2 * n
```

The image of the function is the set of even integers; this is a proper subset of the result type, so times_two is not surjective.

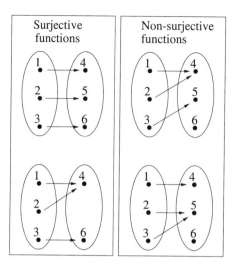

Figure 9.8: Two Surjective Functions and Two that are Not Surjective

Example 115. The increment function takes an integer argument and adds 1 to it. The result type is Integer, and the image is also Integer, since every integer is 1 greater than its predecessor. Thus the image set is the same as the result type, and the function is surjective.

Example 116. Let $A = \{1, 2, 3\}$ and $B = \{4, 5\}$. Define $f :: A \rightarrow B$ as $\{(1, 4), (2, 5), (3, 4)\}$. Then f is surjective, since there is an ordered pair whose second component is 4, and the same is also true of 5.

Example 117. Figure 9.8 shows two surjective functions and two that are not surjective. The domain and image of each function is circled.

   If the result type of a function is larger than its domain, then the function cannot be surjective.

Example 118. Let $A = \{2, 3\}$ and $B = \{4, 5, 6\}$. If $f :: A \rightarrow B$ then it is not surjective, since it is not possible for it to contain three ordered pairs $(x_1, 4)$, $(x_2, 5)$ and $(x_3, 6)$ such that x_1, x_2 and x_3 are all elements of A and are all distinct.

Example 119. The following Haskell functions are surjective:

```
not :: Bool -> Bool
member v :: [Int] -> Bool
increment :: Int -> Int
id :: a -> a
```

   All of these functions have result types that are no larger than their domain types.

Example 120. The following Haskell functions are not surjective. The `length` function returns only zero or a positive integer, and `abs` returns the absolute value of its argument, so both functions can never return a negative number. The `times_two` function may be applied to an odd or an even number, but it always returns an even result.

```
length :: [a] -> Int
abs :: Int -> Int
times_two :: Int -> Int
```

The software tools file defines the following function, which takes a graph representation of a function and determines whether it is surjective:

```
isSurjective
   :: (Eq a, Show a, Eq b, Show b)
   => Set a -> Set b -> Set (a,FunVals b) -> Bool
```

Exercise 9. Decide whether the functions represented by the graphs in the following examples are surjective, and then check using the computer:

```
isSurjective [1,2,3] [4,5]
               [(1, Value 4), (2, Value 5), (3, Value 4)]

isSurjective [1,2,3] [4,5]
               [(1, Value 4), (2, Value 4), (3, Value 4)]
```

Exercise 10. Which of the following functions are surjective?

(a) $f :: A \to B$, where $A = \{1, 2\}$, $B = \{2, 3, 4\}$ and $f = \{(1, 2), (2, 3)\}$.

(b) $g :: C \to D$, where $C = \{1, 2, 3\}$, $D = \{1, 2\}$ and $g = \{(1, 1), (2, 1), (3, 2)\}$.

Exercise 11. Which of the following functions are *not* surjective, and why?

(a) `map increment ::    [Int] -> [Int]`

(b) `take 0 ::    [a] -> [a]`

(c) `drop 0 ::    [a] -> [a]`

(d) `(++) xs ::    [a] -> [a]`

9.6.2 Injective Functions

The essential requirement that a relation must satisfy in order to be a function is that it maps each element of its domain to *exactly one* element of the image. An injective function has a similar property: each element of the image is the result of applying the function to *exactly one* element of the domain.

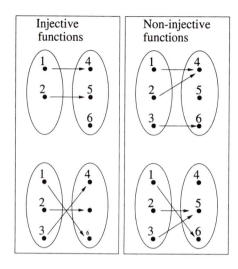

Figure 9.9: Two Injective Functions and Two that are Not Injective

Definition 65. The function $f :: A \to B$ is *injective* if

$$\forall a, a' \in A.\ a \neq a' \to f\ a \neq f\ a'$$

Example 121. Let $A = \{1, 2\}$ and $B = \{3, 4, 5\}$. Define $f :: A \to B$ as $\{(1, 3), (2, 5)\}$. Then f is injective, because 3 appears in only one ordered pair, $(1, 3)$, and 5 appears in only one ordered pair, $(2, 5)$.

Example 122. Let $A = \{1, 2, 3\}$ and $B = \{4, 5\}$. Define $g :: A \to B$ as $\{(1, 4), (2, 5), (3, 5)\}$. Then g is not injective because it contains the ordered pairs $(2, 5)$ and $(3, 5)$, which have different argument values but the same result value.

Example 123. Figure 9.9 shows four functions, two of which are injective.

Example 124. The following Haskell functions are injective:

```
(/\) True :: Bool -> Bool
increment :: Int -> Int
id :: a -> a
times_two :: Int -> Int
```

Example 125. The following Haskell functions are not injective. The `length` function is not injective, because it will map both `[1,2,3]` and `[3,2,1]` to the same number, 3. There are infinitely many other examples that would suffice to show that `length` is not injective, but you only have to give one. The function `take n` is not injective because it will map both $[a_1, a_2, \ldots, a_n, x]$ and $[a_1, a_2, \ldots, a_n, y]$ to the same result $[a_1, a_2, \ldots, a_n]$, even if $x \neq y$.

```
length :: [a] -> Int
take n :: [a] -> [a]
```

The software tools file defines the following function, which determines whether a function (specified by its graph) is injective. The first argument is the function's domain, represented as a set; the second argument is its result type, also represented as a set; and the third argument is its graph.

```
isInjective
    :: (Eq a, Show a, Eq b, Show b)
    => Set a -> Set b -> Set (a, FunVals b) -> Bool
```

Example 126. Let $A = \{1, 2, 3\}$ and $B = \{4, 5, 6\}$. Define three functions as follows:

$$
\begin{aligned}
f_1, f_2, f_3 &\ ::\ A \to B \\
f_1 &= \{(1,4), (2,6), (3,5)\} \\
f_2 &= \{(1,4), (2,4), (3,5)\} \\
f_3 &= \{(1,4), (3,5)\}
\end{aligned}
$$

We can use the software tools to explore the properties of various compositions of these functions. First we need to represent the function graphs in Haskell:

```
fun_domain = [1,2,3]
fun_codomain = [4,5,6]

fun1 = [(1, Value 4), (2, Value 6), (3, Value 5)]
fun2 = [(1, Value 4), (2, Value 4), (3, Value 5)]
fun3 = [(1, Value 4), (2, Undefined), (3, Value 5)]
```

Now we can try various experiments:

```
isInjective  fun_domain fun_codomain
    (functionalComposition fun1 fun2)
isInjective  fun_domain fun_codomain
    (functionalComposition fun1 fun3)
isInjective  fun_domain fun_codomain
    (functionalComposition fun2 fun3)
```

Exercise 12. Determine whether the functions in these examples are injective, and check your conclusions using the computer:

    (a)   `isInjective [1,2,3] [2,4]`
                         `[(1,Value 2),(2,Value 4),(3,Value 2)]`

    (b)   `isInjective [1,2,3] [2,3,4]`
                         `[(1,Value 2),(2,Value 4),(3,Undefined)]`

Exercise 13. Which of the following functions are injective?

(a) $f :: A \rightarrow B$, where $A = \{1, 2\}$, $B = \{1, 2, 3\}$ and $f = \{(1, 2), (2, 3)\}$.

(b) $g :: C \rightarrow D$, where $C = \{1, 2, 3\}$, $D = \{1, 2\}$ and $g = \{(1, 1), (2, 2)\}$.

Exercise 14. Suppose that $f :: A \rightarrow B$ and A has more elements than B. Can f be injective?

9.6.3   The Pigeonhole Principle

The Pigeonhole Principle is a common-sense form of reasoning about the relationship between two finite sets. It says that if A and B are finite sets, where $|A| > |B|$ then no injection exists from A to B. In other words, since each element of the domain must be assigned a pigeonhole in the image, and the domain is bigger than the image, then there must be an element left over, *since at most one pigeon fits in a pigeonhole*. This principle is frequently used in set theory proofs, especially proofs of theorems about functions. There are a variety of ways in which to state the pigeonhole principle formally.

Theorem 68 (Pigeonhole Principle). Let A and B be finite sets, such that $|A| > |B|$ and $|A| > 1$. Let $f :: A \rightarrow B$. Then

$$\exists a_1, a_2 \in A. \, (a_1 \neq a_2) \wedge (f \, a_1 \, = \, f \, a_2).$$

9.7   Bijective Functions

Definition 66. A function is *bijective* if it is both surjective and injective. An alternative name for 'bijective' is *one-to-one and onto*. A bijective function is sometimes called a *one-to-one correspondence*.

Example 127. Figure 9.10 shows some bijective functions and some that are not bijective.

The domain and image of a bijective function must have the same number of elements. This is stated formally in the following theorem:

Theorem 69. Let $f :: A \rightarrow B$ be a bijective function. Then $|\text{domain } f| = |\text{image } f|$.

Proof. Suppose that the domain A is larger than the image B. Then f cannot be injective, by the Pigeonhole Principle. Now suppose that B is larger than A. Then not every element of B can be paired with an element of A: there are too many of them, so f cannot be surjective. Thus a function is bijective only when its domain and image are the same size.   □

A bijective function must have a domain and image that are the same size, and it must also be surjective and injective. As before, we assume that

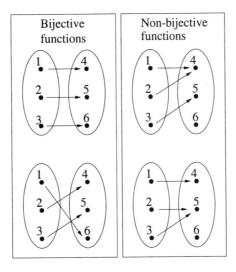

Figure 9.10: Two Bijective Functions and Two that are Not Bijective

these functions are finite. The following function, defined in the software tools file, takes a domain, codomain and a function, and it determines whether the function is bijective.

```
isBijective
   :: (Eq a, Show a, Eq b, Show b)
   => Set a -> Set b
      -> Set (a,FunVals b) -> Bool
```

Exercise 15. Determine whether the following functions are bijective, and check your conclusions using the computer:

```
isBijective [1,2] [3,4] [(1,Value 3),(2,Value 4)]
isBijective [1,2] [3,4] [(1,Value 3),(2,Value 3)]
```

9.7.1 Permutations

Definition 67. A *permutation* is a bijective function $f :: A \to A$; i.e. it must have the same domain and image.

Example 128. The identity function is a permutation.

The only thing a permutation function can do is to shuffle its input; it cannot produce any results that do not appear in its input.

Example 129. Let $A = \{1, 2, 3\}$ and let $f :: A \to A$ be defined by the graph $\{(1, 2), (2, 3), (3, 1)\}$. Then f is a permutation.

Example 130. Let $X = [x_1, x_2, \ldots, x_n]$ be an array of values, and let $Y = [y_1, y_2, \ldots, y_n]$ be the result of sorting X into ascending order. Define $A = \{1, 2, \ldots, n\}$ to be the set of indices of the arrays X and Y. We can define a function $f :: A \to A$ which takes the index of a data value in X and returns the location of that same data value in Y. Then f is a permutation.

Sometimes it is convenient to think of a permutation as a function that reorders the elements of a list; this is often simpler and more direct than defining a function on the indices. For example, it is natural to say that a sorting function has type $[a] \to [a]$. Technically, the function f used in Example 130 is a permutation. The following definition provides a convenient way to represent a permutation as a function that reorders the elements of a list:

Definition 68. A *list permutation function* is a function $f :: [a] \to [a]$ which takes a list of values and rearranges them using a permutation g, such that

$$xs \,!!\, i = (f \ xs) \,!!\, (g \ i).$$

Example 131. The functions `sort` and `reverse` are list permutation functions.

If you rearrange a list of values and then rearrange them again, you simply end up with a new rearrangement of the original list, and that could be described directly as a permutation. This idea is stated formally as follows:

Theorem 70. Let $f, g :: A \to A$ be permutations. Then their composition $f \circ g$ is also a permutation.

The following function, defined in the software tools file, determines whether a function is a permutation. The first two arguments are the domain and result type, which must be finite sets with equality, and the third argument is the function graph.

```
isPermutation
  :: (Eq a, Show a)
  => Set a -> Set a
     -> Set (a,FunVals a) -> Bool
```

Exercise 16. Let $A = \{1, 2, 3\}$ and $f :: A \to A$, where
$f = \{(1,3), (2,1), (3,2)\}$. Is f bijective? Is it a permutation?

Exercise 17. Determine whether the following functions are permutations, and check using the computer:

```
isPermutation
  [1,2,3] [1,2,3]
  [(1,Value 2),(2, Value 3),(3, Undefined)]
isPermutation
  [1,2,3] [1,2,3]
  [(1,Value 2),(2, Value 3),(3, Value 1)]
```

Exercise 18. Is f, defined below, a permutation?

```
f :: Integer -> Integer
f x = x+1
```

Exercise 19. Suppose we know that the composition $f \circ g$ of the functions f and g is surjective. Show that f is surjective.

9.7.2   Inverse Functions

A function $f :: A \to B$ takes an argument $x :: A$ and gives the result $(f\ x) :: B$. The inverse of the function goes the opposite direction: given a result $y :: B$, it produces the argument $x :: A$ which would cause f to yield the result y.

Not all functions have an inverse. For example, if both $(1,5)$ and $(2,5)$ are in the graph of a function, then there is no unique argument that yields 5. Therefore the definition of inverse requires the function to be a bijection.

Definition 69. Let $f :: A \to B$ be a bijection. Then the *inverse* of f, denoted f^{-1}, has type $f^{-1} :: B \to A$, and its graph is

$$\{(y, x) \mid \exists x, y.\ (x, y) \in f\}.$$

Example 132. Let $A = \{1, 2, 3\}$, $B = \{4, 5, 6\}$ and let $f :: A \to A$ have the graph
$\{(1, 4), (2, 5), (3, 6)\}$. Then its inverse is $f^{-1} = \{(4, 1), (5, 2), (6, 3)\}$.

Example 133. The Haskell function decrement is the inverse of increment. Similarly, increment is the inverse of decrement.

```
increment, decrement :: Integer -> Integer
increment x = x+1
decrement x = x-1
```

Exercise 20. Suppose that $f :: A \to A$ is a permutation. What can you say about f^{-1}?

9.8   Cardinality of Sets

One of the most fundamental properties of a set is its size, and this must be defined carefully because of the subtleties of infinite sets. Bijections are a crucial tool for reasoning about the sizes of sets.

A bijection is often called a one-to-one correspondence, which is a good description: if there is a bijection $f :: A \to B$, then it is possible to associate each element of A with exactly one element of B, and vice versa. This is really what you are doing when you count a set of objects: you associate 1 with one of them, 2 with the second, and so on. When you have associated all the objects with a number, then the number n associated with the last one is the number of objects. Thus the number of objects in S is n, if there is a bijection $f :: \{1, 2, \ldots, n\} \to S$. This idea is used formally to define the size of a set, which is called its *cardinality*:

Definition 70. A set S is *finite* if and only if there is a natural number n such that there is a bijection mapping the natural numbers $\{0, 1, \ldots, n - 1\}$ to S. The *cardinality* of S is n, and it is written as $|S|$.

In other words, if S is finite then it can be counted, and the result of the count is its cardinality (i.e. the number of elements it contains).

We would also like to define what it means to say that a set is infinite. It would be meaningless to say 'a set is infinite if the number of elements is infinity', because infinity is not a natural number. We need to find a more fundamental property that distinguishes infinite sets from finite ones.

A relevant observation is that we can make a one-to-one correspondence between the set N of natural numbers and the set E of even numbers, and yet there are natural numbers which are not even.

$$0 \quad 1 \quad 2 \quad 3 \quad 4 \quad \ldots$$
$$0 \quad 2 \quad 4 \quad 6 \quad 8 \quad \ldots$$

Now, we can calculate the ith element of the second row by applying f to the ith element of the first row, where $f :: N \to E$ is defined by $f\ x\ =\ 2 \times x$. Furthermore, f is an injective function, and E is a *proper* subset of N. It would certainly be impossible to find an f with these properties for a finite set. This suggests a method for defining infinite sets:

Definition 71. A set A is *infinite* if there exists an injective function $f :: A \to B$ such that B is a proper subset of A.

We can use the properties of a function over a finite domain A and result type B to determine their relative cardinalities:

- If f is a surjection then $|A| \geq |B|$.

- If f is an injection then $|A| \leq |B|$.

- If f is a bijection then $|A| = |B|$.

Earlier we discussed counting the elements of a finite set, placing its elements in a one-to-one correspondence with the elements of $\{1, 2, \ldots, n\}$. Even though there is no natural number n which is the size of an infinite set, we can use a similar idea to define what it means to say that two sets have the same size, *even if they are infinite*:

Definition 72. Two sets A and B have the same cardinality if there is a bijection $f :: A \to B$.

Example 134. Let $A = \{1, 2, 3\}$ and $B = \{\text{cat}, \text{mouse}, \text{rabbit}\}$. Define $f ::$ $A \rightarrow B$ as

$$f = \{(1, \text{cat}), (2, \text{mouse}), (3, \text{rabbit})\}.$$

Now f is surjective and injective (you should check this), so it is a bijection. Hence the cardinality of B is

$$|B| = 3.$$

The previous example may look unduly complicated, but the point is that exactly the same technique can be used to investigate the sizes of infinite sets.

Example 135. We can place the set I of integers into one-to-one correspondence with the set N of natural numbers:

$$
\begin{array}{ccccccccc}
N = & 0 & 1 & 2 & 3 & 4 & 5 & 6 & \cdots \\
I = & 0 & -1 & 1 & -2 & 2 & -3 & 3 & \cdots
\end{array}
$$

This is done with the function $f :: I \rightarrow N$, defined as:

$$
fx = \begin{cases} 2 \times x, & \text{if } x \geq 0 \\ -2 \times x - 1, & \text{if } x < 0 \end{cases}
$$

Now f is a bijection (you should check that it is), so I has the same cardinality as N.

We have already established that the cardinality of the set of even numbers is the same as the cardinality of N, so this is also the same as the cardinality of I. The size of the set of integers is the same as the size of the set of integers that are non-negative and even!

Definition 73. A set S is *countable* if and only if there is a bijection $f :: N \rightarrow$ S.

A set is countable if it has the same cardinality as the set of natural numbers. In daily life, we use the word *counting* to describe the process of enumerating a set with $1, 2, 3, \ldots$, so it is natural to call a set countable if it can be enumerated—even if the set is infinite.

Exercise 21. Explain why there cannot be a finite set that satisfies Definition 71.

Exercise 22. Suppose that your manager gave you the task of writing a program that determined whether an arbitrary set was finite or infinite. Would you accept it? Explain why or why not.

Exercise 23. Suppose that your manager asked you to write a program that decided whether a function was a bijection. How would you respond?

9.8.1 The Rational Numbers are Countable

It turns out that some infinite sets are countable and others are not, as we will see shortly. In computer science applications it is particularly important to know whether an infinite set is countable, since a computer performs a sequence of operations which is countable. It is often possible for a computer to print out the elements of a countable set; it will never finish printing the entire set, but any specific element will eventually be printed. However, if a set is not countable, then the computer will not even be able to ensure that a given element will eventually be printed.

A rational number is a fraction of the form x/y, where x and y are integers. We can represent a ratio as a pair of integers, the first being the numerator and the second the denominator.

Now suppose that we want to enumerate all of these ratios: we need to put them into one-to-one correspondence with N. Our goal is to create a series of columns, each of which has an index n indicating its place in the series. A column gives all possible fractions with n as the numerator.

$$(1,1)$$
$$(1,2) \quad (2,1)$$
$$(1,3) \quad (2,2) \quad (3,1)$$
$$(1,4) \quad (2,3) \quad (3,2) \quad (4,1)$$
$$(1,5) \quad (2,4) \quad (3,3) \quad (4,2) \quad (5,1)$$
$$\vdots \qquad \vdots \qquad \vdots \qquad \vdots \qquad \vdots$$

Every line in this sequence is finite, so it can be printed completely before the next line is started. Each time a line is printed, progress is made on all of the columns and a new one is added. Every ratio will eventually appear in the enumeration. Thus the set Q of rational numbers can be placed in one-to-one correspondence with N, and Q is countable.

Exercise 24. The software tools file contains a definition of a list named `rationals`, which uses the enumeration illustrated above. Try evaluating the following expressions with the computer:

```
take 3 rationals
take 15 rationals
```

9.8.2 The Real Numbers are Uncountable

Obviously, if $A \subseteq B$ we cannot say that set B is larger than set A, since they might be equal. Surprisingly, even if we know that A is a *proper* subset of B, $A \subset B$, we *still* cannot say that B has more elements: as the previous section demonstrated, it is possible that both are infinite but countable. For example, the set of even numbers is a proper subset of the set of natural numbers, yet both sets have the same cardinality!

It turns out that some infinite sets are not countable. Such a set is so much bigger than the set of natural numbers that there is no possible way to make a one-to-one correspondence between it and the naturals. The set of real numbers has this property. In this section we will explain why, and introduce a technique called *diagonalisation* which is useful for showing that one infinite set has a larger cardinality than another.

We will not give a formal definition of the set of real numbers, but will simply consider a real number to be a string of digits. Consider just the real numbers x such that $0 \le x < 1$; these can be written in the form $.d_0 d_1 d_2 d_3 \ldots$. There is no limit to the length of this string of digits.

Now, suppose that there is some clever way to place the set of real numbers into a one-to-one correspondence with the set of natural numbers (that is, suppose the set of reals is countable). Then we can make a table, where the ith row contains the ith real number x_i, and it contains the list of digits comprising x_i. Let us name the digits in that list $d_{i,0} d_{i,1} d_{i,2} \ldots$. Thus $d_{i,j}$ means the jth digit in the decimal representation of the ith real number x_i. Here, then, is the table which—it is alleged—contains a complete enumeration of the set of real numbers:

$$
\begin{array}{ccccc}
.d_{00} & d_{01} & d_{02} & d_{03} & \cdots \\
.d_{10} & d_{11} & d_{12} & d_{13} & \cdots \\
.d_{20} & d_{21} & d_{22} & d_{23} & \cdots \\
\cdots & \cdots & \cdots & \cdots & \cdots
\end{array}
$$

Now we are going to show that this list is incomplete by constructing a new real number y which is definitely not in the list. This number y also has a decimal representation, which we will call $.d_{y0} d_{y1} d_{y2} \ldots$. Now we have to ensure that y is different from x_0, and it is sufficient to make the 0th digit of y (i.e. d_{y0}) different from the corresponding digit of x_0 (i.e. d_{00}). We can do this by defining a function *different* $:: Digit \to Digit$. There are many ways to define this function; here is one:

$$
\textit{different } x = \begin{cases} 0, & \text{if } x \ne 0 \\ 1, & \text{if } x = 0 \end{cases}
$$

It doesn't matter exactly how *different* is defined, as long as it returns a digit which is different from its argument.

We need to ensure that y is different from x_i for *every* $i \in N$, not just for x_0. This is straightforward: just make the ith digit of y different from the ith digit of x_i:

$$
d_{yi} = \textit{different } x_{ii}.
$$

Now we have defined a new number y; it is real, since it is defined by a sequence of digits, and it is different from x_i for any i. Furthermore, our construction of y did not depend on knowing how the enumeration of x_i worked:

for *any alleged enumeration whatsoever* of the real numbers, our construction will give a new real number which is not in that list. The conclusion, therefore, is that it is impossible to set up a one-to-one correspondence between the set R of reals and the set N of naturals. R is infinite and uncountable.

9.9   Suggestions for Further Reading

The books on set theory cited in Chapter 4 also explain the basic properties of functions.

In Section 9.4 we proved that the Halting Problem is unsolvable. This result has connections to undecidability (see the readings suggested for Chapter 2) and it is fundamental to computability theory, which is covered in many standard textbooks.

An interesting class of function is cryptography algorithms. These range from simple ciphers, which provide interesting applications of the basic properties of functions, all the way to modern public key systems. A history of the subject is given by Singh [25].

9.10   Review Exercises

Exercise 25. A program contains the expression (f.g) x.

(a) Suppose that when this is evaluated, the g function goes into an infinite loop. Does this mean that the entire expression is $\perp$?

(b) Now, suppose that the application of f goes into an infinite loop. Does this mean that the entire expression is $\perp$?

Exercise 26. Each part of this exercise is a statement that might be *correct or incorrect*. Write Haskell programs to help you experiment, so that you can find the answer.

(a) Let $f \circ g$ be a function. If f and g are surjective then $f \circ g$ is surjective.

(b) Let $f \circ g$ be a function. If f and g are injective then $f \circ g$ is injective.

(c) If $f \circ g$ is bijective then f is surjective and g is injective.

(d) If f and g are bijective then $f \circ g$ is bijective.

Exercise 27. The argument and result types given here are sets, not expressions or types in Haskell. Given the functions

```
f : {1,2,3} -> {4,5,6}
f 1 = 4
f 2 = 6
f 3 = 5
```

```
g : {4,5,6} -> {1,2,3}
g 4 = 1
g 5 = 1
g 6 = 2
```

what is

```
(g o f) 1
(g o f) 3
(f o g) 4
(f o g) 5
```

Exercise 28. State the properties of the following functions:

```
f : {3,4,5} -> {3,4,5}
f 3 = 4
f 4 = 5
f 5 = 3

g : {0,1,2} -> {0,1,2}
g 0 = 0
g 1 = 1
g 2 = 2

h : {3,4,5} -> {3,4,5}
h 4 = 3
h 5 = 4
h 3 = 5
```

Exercise 29. Given the functions

```
f : {x,y,z} -> {7,8,9,10}
f x = 8
f y = 10
f z = 7

g : {7,8,9,10} -> {x,y,z}
g 7 = x
g 8 = x
g 9 = x
g 10 = x

h : {7,8,9,10} -> {7,8,9,10}
h 7 = 10
h 8 = 7
h 9 = 8
h 10 = 9
```

describe the following functions:

```
g o f
h o f
g o h
```

Exercise 30. Given the domain and codomain $\{1, 2, 3, 4, 5\}$, which of the following are functions?

```
f 1 = 2
f 2 = 3
f 3 = 3
f 3 = 4
f 4 = 4
f 5 = 5

g 1 = 2
g 2 = 1
g 3 = 4
g 4 = 4
g 5 = 3

h 1 = 2
h 2 = 3
h 3 = 4
h 4 = 1
```

Exercise 31. Determine which of the following definitions are partial functions over the set $\{1, 2, 3\}$.

```
f 1 = undefined
f 2 = 1
f 3 = 2

g 1 = 3
g 2 = 2
g 3 = 1

h 1 = undefined
h 2 = undefined
h 3 = undefined
```

Exercise 32. The following functions are defined over the sets $\{1, 2, 3\}$ and $\{7, 8, 9, 10\}$.

```
f 1 = 7
f 2 = 8
```

```
f 3 = 9

g 7 = 1
g 8 = 2
g 9 = 3
g 10 = 1

h 1 = 3
h 2 = 2
h 3 = 1
```

which of the following are surjections?

```
h o h
f o g
g o f
h o f
g o h
```

Exercise 33. The functions f, g and h are defined over the sets $\{1, 2, 3\}$ and $\{4, 5, 6\}$; which of them are injections?

```
f 1 = 4
f 2 = 5
f 3 = 5

g 4 = 1
g 5 = 2
g 6 = 3

h 4 = 1
h 5 = 1
h 6 = 1
```

Exercise 34. Consider the following functions defined over the sets $\{1, 2, 3\}$ and $\{6, 7, 8, 9\}$; which of them are bijections?

```
f 6 = 1
f 7 = 2
f 8 = 3
f 9 = 3

g 1 = 3
g 2 = 2
g 3 = 1

h 1 = 6
```

```
h 2 = 7
h 3 = 8

g o g
h o f
f o h
```

Exercise 35. Which of these functions is a partial function?

```
function1 True = False
function1 False = function1 False

function2 True = True
function2 False = True
```

Exercise 36. Using `normalForm` and `map`, write a function that takes a list of pairs and determines whether the list represents a function. You can assume in this and the following questions that the domain is the set of first elements of the pairs and the image is the set of second pair elements.

Exercise 37. Using `normalForm` and `map`, define a function `isInjection` so that it returns True if the argument represents an injective function and False otherwise.

Exercise 38. Is it possible to write a function that determines whether a list of pairs represents a surjective function *without* passing in the codomain of the function?

Exercise 39. How much information would you need to know about a Haskell function in order to be able to tell that it is not the identity function?

Exercise 40. Write a function with type

```
compare
  :: (Eq a, Eq b, Eq c, Show a, Show b, Show c)
  => (a -> b)
     -> (b -> c) -> (a -> c) -> a -> Bool
```

that takes three functions f, g and h and determines whether f o g = h for some value of type a.

Exercise 41. Is this definition of `isEven` inductive?

```
isEven :: Int -> Bool
isEven 0 = True
isEven 1 = False
isEven n = isEven (n-2)
```

Exercise 42. Is this definition of isOdd inductive?

```
isOdd 0 = False
isOdd 1 = True
isOdd n =
  if (n < 0) then isOdd (n+2) else isOdd (n-2)
```

Chapter 10

Discrete Mathematics in Circuit Design

The techniques of discrete mathematics which you have been studying in this book are used throughout computer science. So far we have seen many small examples of the application of mathematics to computing, and we have also used programming to help with the mathematics.

This chapter is a complete change of pace: there won't be any new mathematics, but instead we explore in depth one typical and important application to computing: the use of discrete mathematics to help with the process of designing digital circuits. The aim of this chapter is to show a substantive example of the practical application of discrete mathematics. In order to do this, it will be necessary to go into some depth in the subject of digital circuits, but hardware design is not the real subject and we do not give a thorough presentation of it here.

In addition to applying discrete mathematics, we will use Haskell to specify and simulate circuits. The combination of discrete mathematics and Haskell makes it possible to carry out several useful tasks: precise specification of circuits, simulation, correctness proofs, and circuit derivations.

Digital circuit design is a vast subject area, and there is not space here to cover all of it. Therefore we will consider only one class of digital circuits (*combinational* circuits, which don't contain flip flops). However, that restriction is made *only* to keep the chapter short; discrete mathematics is used heavily throughout the entire subjects of digital circuit design and computer architecture.

You do not need to have any prior knowledge about hardware in order to read this chapter; everything you need to know is covered here. We will begin by defining the basic hardware components, Boolean logic gates, and then will look at how to specify and simulate simple circuits using Haskell. Then we apply the methods of Propositional Logic to circuit design, including reasoning

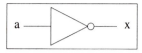

Figure 10.1: Symbol for the Inverter

with truth tables and algebraic reasoning about circuits.

One of the principal problems in designing circuits is ensuring that they are correct. It is extremely expensive to debug circuit designs by building and testing them, so getting the design right in the first place is crucial. The last sections in this chapter address this issue: we use mathematics to state precisely what it means for an addition circuit to be correct. We use recursion and higher order functions to design an adder, and we use induction to prove that it works correctly. In the course of solving this eminently practical problem, we will use more than half of the mathematical topics covered in this book.

10.1   Boolean Logic Gates

Digital circuits are constructed with primitive circuits called *logic gates*. There are logic gate that implement the basic operations of propositional logic, $\wedge$, $\vee$, $\neg$, as well as a few other similar operations.

The simplest logic gate is the *inverter*, which implements the logical not ($\neg$) operation. It takes one input and produces one output; Figure 10.1 shows the standard symbol for the inverter with input a and output x. Instead of using the $\neg$ symbol to specify an inverter, we will use the name *inv*; thus *inv a* means the output of an inverter whose input is a. The inverter's truth table is identical to the truth table for the logical not ($\neg$) operator. It is traditional in circuit design to use 0 and 1 rather than **False** and **True**.

a	*inv a*
0	1
1	0

Some of the most commonly used logic gates take two inputs. The logical $\wedge$ operation is performed by the *and2* gate, whose symbol is shown in Figure 10.2. This gate is named *and2* because it takes two inputs; there are similar gates *and3*, *and4* that take 3 and 4 inputs respectively. The inclusive logical or operation $\vee$ is produced by the *or2* gate (Figure 10.3), and the exclusive or operation, which produces 1 if either argument is 1 *but not both*, is provided by the *xor* gate (Figure 10.4). The following table defines the outputs of these logic gates.

Figure 10.2: And Gate

Figure 10.3: Or Gate

a	b	$and2\ a\ b$	$or2\ a\ b$	$xor\ a\ b$
0	0	0	0	0
0	1	0	1	1
1	0	0	1	1
1	1	1	1	0

10.2 Functional Circuit Specification

Since a digital circuit produces outputs that depend on its inputs, a circuit can be modelled by a mathematical function. Furthermore, we can implement such functions directly with Haskell functions. This provides several valuable benefits, including error checking, powerful specification techniques and tools for circuit simulation.

A circuit can be specified in two ways: using a Haskell function definition, or using a schematic diagram. Most of the time we will use both forms, and later in the chapter you will see some of the advantages of each kind of specification. Meanwhile, we will look at how to write functional circuit specifications and what the corresponding schematic diagrams look like.

A circuit's function is applied to its inputs, and the result is the output. For example, to specify that a and b should be connected to the inputs of an

Figure 10.4: Exclusive Or Gate

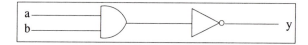

Figure 10.5: The Circuit y = inv (and2 a b)

and2 gate, and the output should be named *x*, we would write the Haskell specification:

```
x = and2 a b
```

Function applications are also used to make connections between several components. The following specification says that the output of the *and2* gate should be inverted, and the inverted output is named *y*. Figure 10.5 shows the corresponding circuit diagram.

```
y = inv (and2 a b)
```

The type of a circuit indicates what its inputs and outputs are. The value carried by a wire is called a *signal*, and there is a class Signal of Haskell types that can be used to represent such a value. Obviously Bool is a member of the Signal class, since we could use True and False to represent the values of logic signals. There are other types as well which are useful in various circumstances. The logic gates have the following types:

```
inv :: Signal a => a -> a
and2, or2, xor :: Signal a => a -> a -> a
```

For every type in the Signal class, there are constants zero and one that represent the basic logic values; these correspond to False and True. In the examples that follow, specialised constants False (for logical 0 or False) and True (for logical 1 or True) will be used; the advantage of False and True is that the system knows which signal type to use for them, so you can omit type signatures when evaluating expressions.

10.2.1   Circuit Simulation

A circuit simulator is a computer program that predicts the behaviour of a circuit, without requiring that the circuit be constructed physically. The program behaves just like the circuit would: it reads in a set of inputs to the circuit, and it produces the same outputs that the real circuit would.

Simulation is important because it is much easier, cheaper and faster to test a design by simulating it with a computer than by constructing it. Just as programs have to be debugged, complex circuit designs also contain errors and must go through an extensive testing and debugging process. With a circuit simulator, it is possible to test the correctness of a design immediately; in contrast, it may take days or weeks to fabricate a physical prototype circuit.

You can simulate any circuit by applying it to suitable signal inputs. For example, we can simulate an and2 gate by applying it to each of the four possible sets of input values(False is 0 and True is 1). Compare the results of the following execution with the truth table for and2:

```
> and2 False False
False
> and2 False True
False
> and2 True False
False
> and2 True True
True
```

A useful technique is to put a set of test cases in the circuit specification file, right after the circuit itself. This serves as documentation, a set of examples to help a reader to understand what is going on, and it's also useful to check that the circuit is still working when other parts of the system have been modified.

Here is a complete set of test cases for simulating the *and2* gate. It is organised just like a truth table: each line consists of a test for particular data values of the inputs. On each line there is a -- symbol which indicates that the rest of the line is a comment, and after the -- we give the expected result.

```
and2 False False    -- False
and2 False True    -- False
and2 True False    -- False
and2 True True    -- True
```

A convenient way to execute the test cases is to put up two windows on the screen: a text editor containing this file, and an interactive session with Haskell. Use the mouse to copy and paste the first line of the test into the Haskell window, and compare the actual result with the expected result. Repeat this for each line in the test suite.

10.2.2 Circuit Synthesis from Truth Tables

Hardware design is not a random process (at least, it shouldn't be). There are many systematic techniques for designing robust circuits. A common situation is that you have the specification in the form of a truth table, and you need to design a circuit which implements that truth table. This section presents a systematic method for solving this problem. It has two great advantages: the method is *simple*, and it *always works*. Sometimes the method doesn't produce the most efficient solution, but that may not be so important, and if it is, there are also systematic methods for optimising circuits.

Every truth table can be written in a general form, where there is one line for every possible combination of input values, and a variable $p, q, r, s \ldots$ specifies the value of the result. For example, here is the general truth table for a circuit f that takes two inputs:

x	y	$f\, x\, y$
0	0	p
0	1	q
1	0	r
1	1	s

A truth table with k input variables will have 2^k lines. To illustrate how to synthesise a logic function, let's consider the following example with three input variables:

a	b	c	$f\, a\, b\, c$
0	0	0	0
0	0	1	1
0	1	0	0
0	1	1	1
1	0	0	1
1	0	1	0
1	1	0	1
1	1	1	0

The idea is simple: the output $f\, a\, b\, c$ should be 1 whenever the inputs correspond to any of the lines where the rightmost column contains 1. All we need is a logical expression which is **True** for each such line and **False** for the others; the value of $f\, a\, b\, c$ is then just the logical or of all those expressions.

Since there happen to be four lines with a result of 1, we will need four expressions, one for each of these lines: we don't yet know what these expressions are, so we just call them $expr_1$, $expr_2$ and so on. The required expression has the form

$$f\, a\, b\, c \;=\; expr_1 \lor expr_2 \lor expr_3 \lor expr_4.$$

The next step is to figure out what these four expressions are. The first one, $expr_1$, should be 1 if the inputs specify the first line of the table where the output is 1. This happens when $a = 0$, $b = 0$ and $c = 1$; equivalently, it happens when $\neg a$, $\neg b$ and c are all true. Therefore the expression is simply $expr_1 = \neg a \land \neg b \land c$. The other expressions are worked out the same way:

a	b	c	x	$expr$
0	0	0	0	
0	0	1	1	$\neg a \wedge \neg b \wedge c$
0	1	0	0	
0	1	1	1	$\neg a \wedge b \wedge c$
1	0	0	1	$a \wedge \neg b \wedge \neg c$
1	0	1	0	
1	1	0	1	$a \wedge b \wedge \neg c$
1	1	1	0	

Now we just plug the expressions into the equation for $f\ a\ b\ c$:

$$f\ a\ b\ c\ =\ (\neg a \wedge \neg b \wedge c) \vee (\neg a \wedge b \wedge c) \vee (a \wedge \neg b \wedge \neg c) \vee (a \wedge b \wedge \neg c)$$

There are several useful refinements of this technique, but those are strictly optional. The important point is that we have a simple method that can be used to synthesise a logical expression—and hence a digital circuit—to implement any truth table.

Often there is a straightforward but inefficient way to design a circuit; an efficient implementation should be used in the final product, but this may be difficult to design. Furthermore, debugging is quicker for easy designs. Because of this, a useful approach is to begin by specifying the simple circuit, and then *transform* it to a more efficient one. The transformation consists of a logical proof that the two circuits have implement exactly the same function. Boolean algebra is a powerful tool for circuit transformation; we can start with a specification expressed as a logical expression, and transform it through a sequence of steps until a circuit with satisfactory performance is found.

Exercise 1. Design a circuit that implements the following truth table:

a	b	c	$f\ a\ b\ c$
0	0	0	1
0	0	1	1
0	1	0	0
0	1	1	0
1	0	0	0
1	0	1	1
1	1	0	1
1	1	1	1

Modern digital circuits can be very large and complex; current processor chips contain several million components. Such circuits cannot be designed as one giant diagram, with every component inserted individually. The key to design is *abstraction*. The circuit is organised in a series of levels of abstraction.

At the lowest level are the logic gates and other primitive components. These are used to design the next level up, including circuits like multiplexors,

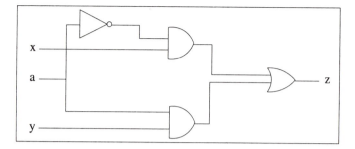

Figure 10.6: Multiplexor

demultiplexors, half and full adders, etc. Such circuits contain around 5 to 10 lower level components, so their specifications are not too complicated. The next level up of the design comprises basic circuits, not primitive logic gates: a design containing 5 or 10 circuits at the level of a multiplexor, for example, would actually contain on the order of 100 logic gates. The same process continues for many levels of abstraction, but at all stages the design is kept reasonably simple through the use of sufficiently high level building blocks.

   This section shows just a few simple circuit designs, in order to give some feeling for how abstraction is used. We will be concerned with just three levels: the primitive logic gates; the simplest circuits, including multiplexors and adders for individual bits, and the next level up where binary numbers are added.

10.2.3   Multiplexors

A multiplexor is the hardware equivalent of a conditional (if—then—else) expression. It takes a control input a and two data inputs, x and y. There is one output; if a is 0 then the output is x, but if a is 1 then the output is y. The circuit is implemented using the standard logic gates (see Figure 10.6):

```
mux1 :: Signal a => a -> a -> a -> a
mux1 a x y = or2 (and2 (inv a) x) (and2 a y)
```

   A demultiplexor is the opposite of a multiplexor. It has a single data input x, and a control input a. The circuit produces two outputs (z_0, z_1). The x input is sent to whichever output is selected by a, and the other output is 0 regardless of the value of x. Figure 10.7 shows the circuit, which is specified as follows:

```
demux1 :: Signal a => a -> a -> (a,a)
demux1 a x = (and2 (inv a) x, and2 a x)
```

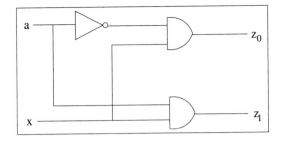

Figure 10.7: Demultiplexor

Figure 10.8: Half Adder Black Box

Exercise 2. Recall the informal description of the multiplexor: if a is 0 then the output is x, but if a is 1 then the output is y. Write a truth table that states this formally, and then use the procedure from Section 10.2.2 to design a multiplexor circuit. Compare your solution with the definition of mux1 given above.

10.2.4 Bit Arithmetic

It's natural to use bits to represent Boolean values and to perform logical calculations with them. An even more common application is to use bits to represent numbers, for example using the binary number system. In fact, the word 'bit' reflects this usage: it originated as an acronym for **Binary Digit**. In this section we will look at digital circuits for adding individual bits; the following sections extend this to words representing binary numbers.

The most basic addition circuit is the *half adder*, which takes two bits a and b to be added together, and produces a two-bit result (c, s) where c is the carry and s is the sum. Figure 10.8 gives the black box diagram for a half adder, and here is its truth table:

a	b	c	s
0	0	0	0
0	1	0	1
1	0	0	1
1	1	1	0

A circuit implementing the half adder could be synthesised using the method given in Section 10.2.2, but we could also just observe that the carry output

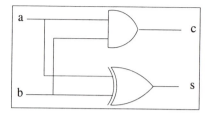

Figure 10.9: Half Adder Circuit

c has the same truth table as the standard *and2* logic gate, while the sum output s is the same as the exclusive or (*xor*). Therefore a half adder can be implemented with just two logic gates (Figure 10.9).

```
halfAdd :: Signal a => a -> a -> (a,a)
halfAdd a b = (and2 a b, xor a b)
```

It is important to be sure that a circuit design is correct before actually fabricating the hardware. One method for improving confidence in correctness is to simulate the circuit. This approach works well for circuits that have a small number of inputs (and no internal state), like `halfAdd`, because it's possible to check every possible combination of input values. Most real-world circuits are too complex for exhaustive testing, and a good approach is to perform some testing and also to carry out a correctness proof. This provides two independent methods for checking the circuit, greatly reducing the likelihood that errors will go unnoticed.

Although `halfAdd` is simple enough to allow complete testing on all possible inputs, we will also consider how to prove its correctness. In order to do this, it's useful to define a *bitValue* function that converts a bit signal into an integer, either 0 or 1. This function requires that the signal arguments be members of the *Static* class, which ensures that they have fixed numeric values. (There are non-static signals used in circuits with flip flops, but those details need not concern us here.)

```
bitValue :: Static a => a -> Int
bitvalue x = if x==zero then 0 else 1
```

The following theorem says that the half adder circuit produces the correct result; that is, if we interpret the output (c, s) as a binary number, then this is actually the sum of the numeric values of the inputs.

Theorem 71. Let $(c, s) = halfAdd\ a\ b$. Then

$$2 \times bitValue\ c + bitValue\ s = bitValue\ a + bitValue\ b.$$

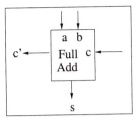

Figure 10.10: Full Adder Black Box

Proof. This theorem is easily proved by checking the equation for each of the four possible combinations of input values. This is essentially the same as using Haskell to simulate the circuit for all the input combinations; the only difference is notational. (Correctness proofs for larger circuits are not essentially the same as simulation.) The details of this proof are left to you to work out. □

In order to add words representing binary numbers, it will be necessary to add three bits: one data bit from each of the words, and a carry input bit. This function is provided by the *full adder* circuit (Figure 10.10); as with the half adder, there is a two-bit result (c', s), where c' is the carry output and s is the sum.

a	b	c	c'	s
0	0	0	0	0
0	0	1	0	1
0	1	0	0	1
0	1	1	1	0
1	0	0	0	1
1	0	1	1	0
1	1	0	1	0
1	1	1	1	1

There are many ways to implement the full adder circuit. The traditional method, given below, uses two half adders. This is an example of the use of abstraction in circuit design: the specification of the full adder is simplified by the use of the `halfAdd` circuit. In general, larger circuits are implemented using a handful of somewhat-smaller circuits, and designers don't implement everything directly using logic gates.

```
fullAdd :: Signal a => (a,a) -> a -> (a,a)
fullAdd (a,b) c = (or2 w y, s)
   where (w,x) = halfAdd a b
         (y,s) = halfAdd x c
```

Figure 10.11: Implementation of Full Adder

The following theorem says that the full adder produces the correct result; it will be needed later to prove the correctness of adders for general binary numbers.

Theorem 72 (Correctness of full adder). Let $(c', s) = fullAdd\ (a, b)\ c$, so that c' is the carry output and s is the sum output. Then

$$bit\,Value\ c' \times 2 + bit\,Value\ s\ =\ bit\,Value\ a + bit\,Value\ b + bit\,Value\ c.$$

Exercise 3. Use Haskell to test the half adder on the following test cases, and check that it produces the correct results.

```
Test cases for half adder, with predicted results
halfAdd False False  -- 0 0
halfAdd False True   -- 0 1
halfAdd True False   -- 1 0
halfAdd True True    -- 1 1
```

Exercise 4. Prove Theorem 71 using truth tables.

10.2.5   Binary Representation

Binary numbers consist of a sequence of bits called a *word*. This is represented as a list. For example, if you have four individual signals named w, x, y and z, you can treat them as a word by writing $[w, x, y, z]$, and its type is *Static* $a \Rightarrow [a]$.

There are, unfortunately, two traditional schemes for numbering the bits in a word: $[x_0, x_1, x_2, x_3]$ and $[x_3, x_2, x_1, x_0]$. We will use the first scheme, where the leftmost bit of a word has index 0 and the rightmost has index $k - 1$, where k is the number of bits in the word. The binary value of the word $[x_0, x_1, x_2, x_3]$ is

$$x_0 \times 2^3 + x_1 \times 2^2 + x_2 \times 2^1 + x_3 \times 2^0.$$

In general, the value of a k-bit word $x = [x_0, \dots, x_{k-1}]$ is

$$\sum_{i=0}^{k-1} x_i \times 2^{k-(i+1)}.$$

Figure 10.12: 4-Bit Ripple Adder Black Box

This value is calculated by the *wordValue* function:

```
wordValue :: Static a => [a] -> Integer
wordValue [] = 0
wordValue (x:xs) = 2^k * bitValue x + wordValue xs
    where k = length xs
```

Notice that in the binary number system the smallest value that can be represented in k bits is 0, and the largest value is $2^k - 1$. Negative numbers are not representable at all in the binary system. Most modern computers represent integers using the *two's complement* number system, which allows for negative numbers. One nice property of two's complement is that an ordinary binary addition circuit can be used to perform addition on two's complement numbers. Consequently we won't worry about negative numbers here, but will proceed to the addition of binary numbers.

Exercise 5. Work out the numeric value of the word $[1, 0, 0, 1, 0]$. Then check your result by using the computer to evaluate:

```
wordValue [True,False,False,True,False]
```

10.3 Ripple Carry Addition

A ripple carry adder (Figure 10.12) is used to calculate the sum of two words. When the word size is four bits, the binary arguments are words containing the bits $[x_0, x_1, x_2, x_3]$ and $[y_0, y_1, y_2, y_3]$. The most significant bits are x_0 and y_0, and appear on the left of the word; the least significant bits are x_3 and y_3, and they appear at the right of the word. The ripple carry adder also takes a carry input c (this makes it possible to add larger numbers by performing a sequence of additions). The output produced by the circuit is a single carry output bit, and a word of sum bits. We require that the two input words and the sum word all contain the same number of bits.

The following specification (Figure 10.13) uses four full adders to construct a 4-bit ripple carry adder. In bit position i, the data inputs are x_i and y_i, and

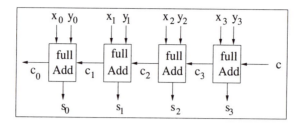

Figure 10.13: 4-Bit Ripple Carry Adder

the carry input is c_{i+1}. The sum produced by position i is s_i, and the carry output c_i will be sent to the bit position to the left (position $i-1$).

```
add4 :: Signal a => a -> [(a,a)] -> (a,[a])
add4 c [(x0,y0),(x1,y1),(x2,y2),(x3,y3)] =
       (c0, [s0,s1,s2,s3])
  where (c0,s0) = fullAdd (x0,y0) c1
        (c1,s1) = fullAdd (x1,y1) c2
        (c2,s2) = fullAdd (x2,y2) c3
        (c3,s3) = fullAdd (x3,y3) c
```

To use the adder, we must convert the input numbers into 4-bit binary representations. For example, here is the addition of $3 + 8$.

```
Example: addition of 3 + 8
   3 + 8
   =   0011    ( 2+1 =   3)
     + 1000    (   8 =   8)
   =   1011    (8+2+1 = 11)
Calculate this by evaluating
   add4 False [(False,True),(False,False),
               (True,False),(True,False)]
The expected result is
   (False, [True,False,True,True])
```

Exercise 6. Use Haskell to evaluate the example above, and check that the result matches the expected result.

10.3.1   Circuit Patterns

The `add4` specification in the previous section is not too complicated, but it would be awfully tedious to extend it to handle words containing 32 or 64 bits (which are the sizes most commonly used with current generation processors).

Figure 10.14: Building Block Circuit for mscanr

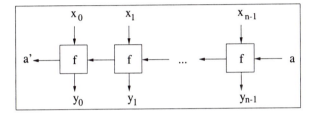

Figure 10.15: Structure of mscanr Pattern

The analogous specifications would contain 32 or 64 local equations, and there would be a correspondingly large number of indexed names. Besides the sheer size, such specifications would be highly error-prone. Furthermore, it would be better to define the family of all ripple carry adders, rather than to keep on defining new ones at various different word sizes.

A much better approach is to define the general k-bit ripple carry adder once and for all, so that it works for arbitrary k. To do this, we can't name the individual bits explicitly, like x_0, x_1 and so on. Instead, we need to use a method that works for any word size without referring explicitly to the individual bits. The most intuitive description of the adder would say 'each full adder has its carry input connected to the carry output of its right neighbour,' and this is exactly the idea that needs to be formalised with a function.

A higher order function can be used to express the abstract structure of circuits like add4. The idea is to write a function whose argument is a circuit specification; the higher order function connects up as many copies as required of the circuit it is given. Figure 10.14 shows the sort of building block needed for the ripple carry adder; it matches the black box structure of the full adder.

The *mscanr* function takes a building-block circuit with an appropriate type. It creates as many copies of the building block as are required, and makes all the internal connections that are needed. Figure 10.15 depicts the structure of the resulting circuit, and Figure 10.16 shows the circuit defined by *mscanr* as a black box.

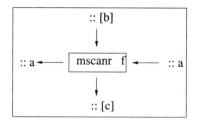

Figure 10.16: Black Box for `mscanr` Pattern

```
mscanr :: (b->a->(a,c)) -> a -> [b] -> (a,[c])
mscanr f a [] = (a,[])
mscanr f a (x:xs) =
  let (a',ys) = mscanr f a xs
      (a'',y) = f x a'
  in (a'', y:ys)
```

Exercise 7. Let $f :: b \to a \to (a, c)$ be a black box circuit. Draw a diagram showing the structure of the circuit specified by

$$mscanr \ f \ a \ [(x_0, y_0), (x_1, y_1), (x_2, y_2)].$$

10.3.2   The n-Bit Ripple Carry Adder

Now we can use the higher order function *mscanr* to define a general ripple carry adder that works for any word size. The definition is much more intuitive than brute-force definitions like *add4*, once you understand the idea of using higher order functions to express regular circuit patterns.

The *mscanr* function expresses the pattern of the ripple carry adder. It simply says that a ripple carry adder consists of a row of full adders, one for every bit position. The carry input to each full adder is connected to the carry output from the full adder to the right, and the carry input to the rightmost (least significant) bit position is the carry input to the entire addition.

The first argument is a circuit specification with type $b \to a \to (a, c)$. Recall that a full adder has type

$$Signal \ a \Rightarrow (a, a) \to a \to (a, a).$$

This fits the *mscanr* pattern, and a ripple carry adder consists of a row of full adders with the carry input of each connected to the carry output of its right neighbour.

```
rippleAdd :: Signal a => a -> [(a,a)] -> (a, [a])
rippleAdd c zs = mscanr fullAdd c zs
```

This definition works for arbitrary word size. The size of a particular circuit is determined by the size of the input data word. The definition itself doesn't get longer if the words become longer! It's now easy to specify a 6-bit adder, as the following test case demonstrates.

```
Example: addition of 23+11
    23 + 11
  =   010111   (16+4+2+1 = 23)
    + 001011   (   8+2+1 = 11) with carry input = 0
  =   100010   (   32+2 = 34) with carry output = 0
Calculate with the circuit by evaluating
    rippleAdd False [(False,False),(True,False),(False,True),
                     (True,False),(True,True),(True,True)]
The expected result is
    (False, [True,False,False,False,True,False])
```

Exercise 8. Work out a test case using the ripple carry adder to calculate $13+41=54$, using 6-bit words. Test it using the computer.

10.3.3 Correctness of the Ripple Carry Adder

Theorem 73. Let xs and ys be k-bit words, so $xs, ys :: Signal\ a \Rightarrow [a]$. Define $(c, sum) = rippleAdd\ zero\ (zip\ xs\ ys)$; thus $c :: a$ is the carry output and $ss :: [a]$ is the sum word. Then

$$bitValue\ c \times 2^k + wordValue\ ss\ =\ wordValue\ xs + wordValue\ ys.$$

The left hand side of the equation is the numeric value of the output of the ripple carry adder circuit, and the right hand side is the numeric value of its inputs. Thus the equation says that the circuit produces the correct answer.

Proof. Induction over zs. For the base case, $k = 0$, $zs = xs = ys = []$. First we simplify:

$$(c, ss)\ =\ rippleAdd\ zero\ []$$
$$= mscanr\ fullAdd\ zero\ []$$
$$= (zero, [])$$
$$c = zero$$
$$ss = []$$
$$wordValue\ []\ +\ wordValue\ []$$
$$= 0 + 0$$
$$= 0$$
$$bitValue\ c \times 2^k + wordValue\ ss$$
$$= 0 \times 2^0 + 0$$
$$= 0 \times 2^0 + wordValue\ []$$

For the inductive case, let $k = length\ xs = length\ ys$, and assume

$$bit\,Value\ c \times 2^k + word\,Value\ ss = word\,Value\ xs + word\,Value\ ys,$$

where $(c, ss) = rippleAdd\ zero\ (zip\ xs\ ys)$. The aim is to prove that

$$bit\,Value\ c \times 2^{k+1} + word\,Value\ ss =$$
$$word\,Value\ (x : xs) + word\,Value\ (y : ys),$$

where $(c, ss) = rippleAdd\ zero\ (zip\ (x : xs)\ (y : ys))$.
    First we simplify:

$(c, ss) = mscanr\ fullAdd\ zero\ (zip\ (x : xs)\ (y : ys))$
    $= mscanr\ fullAdd\ zero\ ((x, y) : zip\ xs\ ys)$
    $= let\ (c', ss) = mscanr\ fullAdd\ zero\ (zip\ xs\ ys)$
        $= rippleAdd\ zero\ (zip\ xs\ ys)$
    $(c'', s) = fullAdd\ (x, y)\ c'$
    $in\ (c'', s : ss)$

Now the left hand side of the equation can be transformed into the right hand sied, using equational reasoning:

lhs (numeric value of output from the adder)
    $= bit\,Value\ c \times 2^{k+1} + word\,Value\ ss$
    $= bit\,Value\ c'' \times 2^{k+1} + word\,Value\ (s : ss)$
    $= bit\,Value\ c'' \times 2^{k+1} + bit\,Value\ s \times 2^k + word\,Value\ ss$
    $= (bit\,Value\ c'' \times 2 + bit\,Value\ s) \times 2^k + word\,Value\ ss$
    $= (bit\,Value\ x + bit\,Value\ y + bit\,Value\ c') \times 2^k + word\,Value\ ss$
    $= (bit\,Value\ x + bit\,Value\ y) \times 2^k + (bit\,Value\ c') \times 2^k + word\,Value\ ss$
    $= (bit\,Value\ x + bit\,Value\ y) \times 2^k + word\,Value\ xs + word\,Value\ ys$
    $= (bbit\,Value\ x \times 2^k + word\,Value\ xs) + (bit\,Value\ y \times 2^k + word\,Value\ ys)$
    $= word\,Value\ (x : xs) + word\,Value\ (y : ys)$
    $= $ rhs (numeric value of inputs to the adder)

□

10.3.4   Binary Comparison

Comparison of binary numbers is just as important as adding them. It is particularly interesting to consider how to implement a comparison circuit, since this problem has some strong similarities and also some strong differences to the ripple carry adder.

First, let's consider the comparison of two bits. This is analogous to starting out with a half adder. The problem is to design a circuit *halfCmp* that compares two one-bit numbers x and y. The result of the comparison should be a triple of bits of the form (lt, eq, gt), where lt is true if $x < y$, eq is true if $x = y$, and *gt* is true if $x > y$. The problem is pretty simple, since x and y are both just one bit! The type should be

$$halfCmp :: Signal\ a \Rightarrow (a,a) \to (a,a,a),$$

where the input bits x and y are provided as a pair (x,y), and the result triple consists of the result bits $(lt,\ eq,\ gt) :: (a,a,a)$. Finding a solution is straightforward:

```
halfCmp :: Signal a => (a,a) -> (a,a,a)
halfCmp (x,y) =
   (and2 (inv x) y,    -- x<y when x=0,y=1
    inv (xor x y),     -- x=y when x=0,y=0 or x=1,y=1
    and2 x (inv y))    -- x>y when x=1,y=0
```

The next problem to consider is that of designing a ripple comparator that takes two words representing binary numbers (the words must have the same size), and returns a triple of three bits $(lt,\ eq,\ gt)$ which indicate the result of comparing x and y. The meanings of the output bits are just the same as in the previous problem; the only difference is that now the inputs to the comparator are words rather than bits.

Just as you compare two numbers by looking first at the most significant digits, a binary comparison is performed by moving from left to right through the word. Initially we assume the two words are equal; represent this by $(lt,\ eq,\ gt) = (0,1,0)$. If the next bit position has $x = 1$ and $y = 0$ then we know that the final result must be $(0,0,1)$ regardless of any bits to the right; conversely if $x = 0$ and $y = 1$ then the final result must be $(1,0,0)$ regardless of the bit values to the right. However, if x and y have the same value in this bit position, then as far as we know the result is still $(0,1,0)$ but that result might be changed later.

The calculation in each bit position requires the two local bits (that is, for position i we need the ith bit of both of the input words). It also requires the result of the comparison for all the bits to the left. The task is performed by a full comparison circuit, which is analogous to the full adder.

```
fullCmp :: Signal a => (a,a,a) -> (a,a) -> (a,a,a)
fullCmp (lt,eq,gt) (x,y) =
   (or2 lt (and3 eq (inv x) y),    -- <
    and2 eq (inv (xor x y)),       -- =
    or2 gt (and3 eq x (inv y)))    -- >
```

Now we can define the ripple comparison circuit, which compares two binary numbers. Its definition is similar to the ripple carry adder, but there are several differences in the circuit pattern required. In the first place, the information flow is left to right for comparison, rather than the right to left order used in addition. Another difference is that for comparison we are interested only in the final horizontally moving value; this would be analogous to wanting the carry output from an addition, but we do not need a result analogous to the sum bits. The standard *foldl* higher order function specifies exactly the circuit pattern needed here. A final difference is that the comparator just takes the two numbers to be compared; it generates the initial horizontal value locally. In contrast, the ripple carry adder takes a carry input. The reason for that is that many applications of adders, such as the ALU of a computer's processor, use carry inputs to provide the ability to add long numbers comprising several words.

```
rippleCmp :: Signal a => [(a,a)] -> (a,a,a)
rippleCmp z = foldl fullCmp (False,True,False) z
```

Exercise 9. Define a full set of test cases for the circuit *halfCmp*, which compares two bits, and execute them using the computer.

Exercise 10. Define three test cases for the *rippleCmp* circuit, with a word size of three bits, demonstrating each of the three possible results. Run your test cases on the computer.

10.4   Suggestions for Further Reading

The application of discrete mathematics to digital circuit design is a large subject. Most of the publications that address this area are aimed more at researchers than students, so some of the references cited here may be difficult to read, but it's interesting to see *real* applications of discrete mathematics.

A paper on the use of functional programming for specifying circuits is [21]. Unfortunately that paper uses an obsolete functional language, which is less readable than Haskell.

10.5   Review Exercises

Exercise 11. Show that the and4 logic gate, which takes four inputs a, b, c and d and outputs $a \wedge b \wedge c \wedge d$, can be implemented using only and2 gates.

Exercise 12. Work out a complete set of test cases for the full adder, and calculate the expected results. Then simulate the test cases and compare with the predicted results.

Exercise 13. Prove Theorem 72.

Exercise 14. Suppose that a computer has 8 memory locations, with addresses $0, 1, 2, \ldots 7$. Notice that we can represent an address with 3 bits, and the size of the memory is 2^3 locations. We name the address bits

$a_0a_1a_2$, where a_0 is the most significant bit and a_2 is the least significant. When a memory location is accessed, the hardware needs to send a signal to each location, telling it whether it is the one selected by the address $a_0a_1a_2$. Thus there are 8 *select* signals, one for each location, named $s_0, s_1, \ldots, s_7$. Design a circuit that takes as inputs the three address bits $a_0a_1a_2$, and which outputs the select signals $s_0s_1 \ldots s_7$. *Hint: use demultiplexors, arranged in a tree-like structure.* (Note: modern computers have an address size from 32 to 64 bits, allowing for a large number of locations, but a 3-bit address makes this exercise more tractable!)

Exercise 15. Does the definition of *rippleAdd* allow the word size to be 0? If not, what prevents it? If so, what does it mean?

Exercise 16. Does the definition of *rippleAdd* allow the word size to be negative? If not, what prevents it? If so, what does it mean?

Exercise 17. Note that for the half adder and full adder, we did thorough testing—we checked the output of the circuit for every possible input. Note also that we did not do this for the ripple carry adder, where we just tried out a few particular examples. The task: Explain why it is infeasible to do thorough testing of a ripple carry adder circuit, and estimate how long it would take to test all possible input values for the binary adder in a modern processor where the words are 64 bits wide.

Exercise 18. Computer programs sometimes need to perform arithmetic, including additions and comparisons, on big integers consisting of many words. Most computer processor architectures provide hardware support for this, and part of that hardware support consists of the ability to perform an addition where the carry input is supplied externally, and is not assumed to be 0. Explain why the carry input to the *rippleAdd* circuit helps to implement multiword addition, but we don't need an analogous horizontal input to *rippleCmp* for multiword comparisons.

Appendix A

Software Tools
for Discrete Mathematics

The Haskell programming language provides excellent support for mathematical computing in general. This book uses programs in Haskell 98, which is standardised, stable and well-supported. The book also requires a library of definitions that give additional support for the topics of discrete mathematics. The library is called *Software Tools for Discrete Mathematics*, and it consists of a single file called stdm.hs. You need, therefore, two pieces of software to use a computer along with the book:

- An interactive implementation of Haskell;

- The file stdm.hs.

Both items are free, and they run on most major computer platforms. All of the software can be downloaded from the web. The web home page for this book (see Appendix B) contains the stdm.hs file along with full documentation, and it also tells you how to download various implementations of Haskell.

Appendix B

Resources on the Web

Home page for *Discrete Mathematics Using a Computer*. The book's web page contains a variety of useful information, and is an integral part of the book:

- You can download the *Software Tools for Discrete Mathematics*, along with complete documentation;

- There are additional practice problems, solutions and explanations;

- There are up-to-date pointers to many other relevant web pages.

```
http://www.dcs.gla.ac.uk/~jtod/discrete-mathematics/
```

Instructor's Guide for Discrete Mathematics Using a Computer. The Instructor's Guide is entirely online. A password is required to read it; please contact the authors to obtain access. See the book home page for the current contact address.

```
http://www.dcs.gla.ac.uk/~jtod
   /discrete-mathematics/instructors-guide/
```

Home page for Haskell. This page contains complete and current information on the Haskell language, including free (and open source) compilers and interpreters which you can download, up-to-date pointers to the home pages for all the Haskell compilers and interpreters, the official Haskell language definition, the complete specification for the standard libraries, pointers to books and articles on Haskell and functional programming, and more.

```
http://www.haskell.org/
```

Appendix C

Solutions to Selected Exercises

C.1   Introduction to Haskell

3.

```
isA :: Char -> Bool
isA 'a' = True
isA c   = False
```

4.

```
isHello :: String -> Bool
isHello ('h':'e':'l':'l':'o':[]) = True
isHello str                      = False
```

5.

```
removeSpace :: String -> String
removeSpace []        = []
removeSpace (' ':xs) = xs
removeSpace xs        = xs
```

6.

```
toBool :: Int -> Bool
toBool 1     = True
toBool 0     = False
toBool other = error "toBool: funny value"

convert :: [Int] -> [Bool]
convert lst = map toBool lst
```

7.

```
member0 :: String -> Bool
member0 string = or (map (== '0') string)
```

8.

```
addJust :: Maybe Int -> Maybe Int -> Maybe Int
addJust (Just a) (Just b) = Just (a + b)
addJust (Just a) Nothing = Just a
addJust Nothing (Just a) = Just a
addJust Nothing Nothing = Nothing

addMaybe :: [Maybe Int] -> [Maybe Int] -> [Maybe Int]
addMaybe lst1 lst2 = zipWith addJust lst1 lst2
```

9.

```
(** 2) 1 ((** 2) 2 ((** 2) 3 0))

seen 3 (seen 2 (seen 1 (seen 4 False)))

spaces "this" (spaces "is"
    (spaces "a" (spaces "sentence" [])))
```

10.

```
data Metals = Copper | Silver | Gold
              | Tin | Platinum | Bronze
              deriving (Eq, Show)
```

11. The coins can be represented by a list containing one or more elements of the following type:

```
data Coins = OneP Int | TwoP Int | FiveP Int
              | TenP Int | TwentyP Int
              | FiftyP Int | HundredP Int
              deriving (Eq, Show)
```

12.

```
data Universal = BOOL Bool | INT Int | CHAR Char
                 deriving (Eq, Show)
```

13.

```
data Tuples a b c d = Tuple0 | Tuple1 a | Tuple2 a b
                      | Tuple3 a b c | Tuple4 a b c d
                      deriving (Eq, Show)
```

15.

```
showMaybe :: Show a => Maybe a -> String
showMaybe Nothing  = []
showMaybe (Just a) = show a
```

16.

```
bitwiseAnd :: [Int] -> [Int] -> [Int]
bitwiseAnd word1 word2 = zipWith bitAnd word1 word2
```

17. Yes, because there is no way to control the depth of the type represented by the type variable a.

18.

```
[True, False]              "2" ++ "a"
[(3,True), (9,False)]      2 == 3
'a' > 'b'                  [[1],[2],[3]]
```

19. The types of the two paired elements must be the same, according to the type signature given. In addition, the type of the first component of the pair was defined as being of the Num class, while the function f was applied to a pair with a Boolean as its first element.

20. There are two possible constructors in the type, and the function definition handles only one of them. If the other is used, an error occurs.

21.

```
[a | (Just a) <- xs]
```

22.

```
largerThanN :: [Int] -> Int -> [Int]
largerThanN lst n = [e | e <- lst, e > n]
```

23.

```
f :: [Int] -> Int -> [Int]
f lst v = [n | n <- [0..length lst - 1], lst!!n == v]
```

24.

```
[e | e <- [1..20],
     [x | x <- [1..e], x * x == e] == []]
```

25.

```
count :: (Eq a, Num b) => a -> a -> b -> b
count letter x acc
 = if letter == x then acc + 1 else acc

countLetters :: Char -> String -> Int
countLetters c lst = foldr (count c) 0 lst
```

26.

```
remove :: Char -> Char -> [Char] -> [Char]
remove ch x acc
 = if x == ch then acc else x:acc

removeEachLetter :: Char -> [Char] -> [Char]
removeEachLetter ch lst
 = foldr (remove ch) [] lst
```

27.

```
rearrange :: [a] -> a -> [a]
rearrange lst x = x:lst

rev :: [a] -> [a]
rev lst = foldl rearrange [] lst
```

28.

```
takeLast :: Maybe a -> a -> Maybe a
takeLast  Nothing x = Just x
takeLast  (Just y) x = Just x

maybeLast :: [a] -> Maybe a
maybeLast lst = foldl takeLast Nothing lst
```

C.2   Propositional Logic

3.

$$\cfrac{R \quad \cfrac{Q \quad R}{Q \wedge R}\,{\scriptstyle\{\wedge I\}}}{P \wedge (Q \wedge R)}\,{\scriptstyle\{\wedge I\}}$$

4. The proof of y will have a symmetrical shape, but the proof of x will appear triangular, with more inference on the right side than on the left. In the general case, x with 2^n variables will have height proportional to 2^n, since every extra variable will require one extra inference above everything else. In contrast, y with 2^n variables will have height proportional to n.

5.

$$
\cfrac{\cfrac{(P \wedge Q) \wedge R}{\cfrac{P \wedge Q}{P}\{\wedge E_L\}}\{\wedge E_L\} \qquad \cfrac{\cfrac{(P \wedge Q) \wedge R}{P \wedge Q}\{\wedge E_L\}}{Q}\{\wedge E_R\} \qquad \cfrac{\cfrac{(P \wedge Q) \wedge R}{R}\{\wedge E_R\} \quad}{Q \wedge R}\{\wedge I\}}{P \wedge (Q \wedge R)}\{\wedge I\}
$$

6.

$$
\cfrac{\cfrac{\cfrac{P \quad \cfrac{P \quad P \to Q}{Q}\{\to E\}}{P \wedge Q}\{\wedge I\} \qquad P \wedge Q \to R \wedge S}{R \wedge S}\{\to E\}}{S}\{\wedge E_R\}
$$

8.

$$
\cfrac{\cfrac{\boxed{Q} \quad \cfrac{P \quad \boxed{Q}}{P \wedge Q}\{\wedge I\}}{Q \to P \wedge Q}\{\to I\}}
$$

11.

$$
\cfrac{\cfrac{P \quad Q}{P \wedge Q}\{\wedge I\}}{(P \wedge Q) \vee (Q \vee R)}\{\vee I_L\}
$$

12. We prove that True ∧ True → True and then that True → True ∧ True, without translating True into False → False.

$$\frac{\dfrac{\text{True} \wedge \text{True}}{\text{True}}\{\wedge E_R\}}{\text{True} \wedge \text{True} \to \text{True}}\{\to I\}$$

$$\frac{\dfrac{\text{True} \quad \text{True}}{\text{True} \wedge \text{True}}\{\wedge I\}}{\text{True} \to \text{True} \wedge \text{True}}\{\to I\}$$

13. We prove that True ∨ False → True and then that True → False ∨ True.

$$\frac{\dfrac{\text{True}}{\text{True} \vee \text{False}}\{\vee I_L\}}{\text{True} \to \text{True} \vee \text{False}}\{\to I\}$$

16.

- P is represented by P

- $Q \vee FALSE$ is represented by Or Q FALSE

- $Q \to (P \to (P \wedge Q))$ is represented by Imp Q (Imp P (And P Q))

18.

$$
\begin{aligned}
&(P \wedge \text{False}) \vee (Q \wedge \text{True}) \\
&= \text{False} \vee (Q \wedge \text{True}) \qquad\quad \{\wedge \text{ null}\} \\
&= \text{False} \vee Q \qquad\qquad\qquad\quad \{\wedge \text{ identity}\} \\
&= Q \qquad\qquad\qquad\qquad\qquad \{\vee \text{ identity}\}
\end{aligned}
$$

20.

$$(P \wedge ((Q \vee R) \vee Q)) \wedge S$$
$$= S \wedge (P \wedge ((Q \vee R) \vee Q)) \qquad \{\wedge \text{ commutative}\}$$
$$= S \wedge (((Q \vee R) \vee Q) \wedge P) \qquad \{\wedge \text{ commutative}\}$$
$$= S \wedge ((Q \vee (R \vee Q)) \wedge P) \qquad \{\vee \text{ associative}\}$$
$$= S \wedge ((Q \vee (Q \vee R)) \wedge P) \qquad \{\vee \text{ commutative}\}$$
$$= S \wedge (((Q \vee Q) \vee R) \wedge P) \qquad \{\vee \text{ associative}\}$$
$$= S \wedge ((Q \vee R) \wedge P) \qquad \{\vee \text{ idempotent}\}$$
$$= S \wedge ((R \vee Q) \wedge P) \qquad \{\vee \text{ commutative}\}$$

23.

$$
\cfrac{
\cfrac{P \quad Q}{P \wedge Q}\{\wedge I\} \qquad \cfrac{R \quad S}{R \wedge S}\{\wedge I\}
}{
(P \wedge Q) \wedge (R \wedge S)
}\{\wedge I\}
$$

24.

$$
\cfrac{
\cfrac{
\cfrac{\boxed{P \wedge Q}}{P}\{\wedge E_L\} \qquad P \to R
}{
R
}\{\to E\}
}{
P \wedge Q \to R
}\{\to I\}
$$

25. We prove that $\mathsf{True} \vee \mathsf{True} \to \mathsf{True}$ and then that $\mathsf{True} \to \mathsf{True} \vee \mathsf{True}$.

$$
\cfrac{
\mathsf{True} \vee \mathsf{True} \qquad \cfrac{\mathsf{True}}{\mathsf{True}}\{ID\} \qquad \cfrac{\mathsf{True}}{\mathsf{True}}\{ID\}
}{
\cfrac{\mathsf{True}}{\mathsf{True} \vee \mathsf{True} \to \mathsf{True}}\{\to I\}
}\{\vee E\}
$$

$$
\cfrac{
\cfrac{\mathsf{True}}{\mathsf{True} \vee \mathsf{True}}\{\vee I_L\}
}{
\mathsf{True} \to \mathsf{True} \vee \mathsf{True}
}\{\to I\}
$$

27. A list comprehension can generate a truth table for you.

```
logicExprValue1 = [((a,b),logicExpr1 a b) |
                   a <- [False,True],
                   b <- [False,True]
                  ]

logicExprValue2 = [((a,b),logicExpr2 a b) |
                   a <- [False,True],
                   b <- [False,True]
                  ]
```

28.

```
logicExprValue3 = [((a,b,c),logicExpr3 a b c) |
                   a <- [False,True],
                   b <- [False,True],
                   c <- [False,True]
                  ]

logicExprValue4 = [((a,b,c),logicExpr4 a b c) |
                   a <- [False,True],
                   b <- [False,True],
                   c <- [False,True]
                  ]
```

29. There are 11 different variables, so a truth table would have 2048 entries in it. On the other hand, a simplification does not help because the expression cannot be simplified very much. It seems best to use Haskell to calculate a truth table, then gather and report some statistics.

```
gatherStats :: String
gatherStats
  = let expr a b c d e f g h i j k
          = (a /\ b \/ c /\ d) /\
            (e \/ f /\ (g /\ h))
          <=>
            (i ==> j /\ (k ==> i \/ b))
        bools = [False,True]
        table
          = [((a,b,c,d,e,f,g,h,i,j,k),
              expr a b c d e f g h i j k) |
               a <- bools, b <- bools, c <- bools,
               d <- bools, e <- bools, f <- bools,
               g <- bools, h <- bools, i <- bools,
               j <- bools, k <- bools]
```

```
          trues
            = [(vs,val) | (vs,val) <- table, val == True]
     in "There are " ++ show (length trues) ++
            " lines for which the expression is True"
```

30.

```
data Logic = A | B | C
           | And Logic Logic
           | Or  Logic Logic
           | Not Logic
           | Imply Logic Logic
           | Equiv Logic Logic
           deriving (Eq, Show)
```

31.

```
distribute :: Logic -> Logic
distribute (And a (Or b c)) = Or (And a b) (And a c)
distribute (Or a (And b c)) = And (Or a b) (Or a c)

deMorgan :: Logic -> Logic
deMorgan (Not (Or a b)) = And (Not a) (Not b)
deMorgan (Not (And a b)) = Or (Not a) (Not b)
deMorgan (And (Not a) (Not b)) = Not (Or a b)
deMorgan (Or (Not a) (Not b)) = Not (And a b)
```

32.

$$
\begin{aligned}
& C \wedge A \wedge B \vee C \\
& = C \vee C \wedge A \wedge B && \{\vee \text{ commutative}\} \\
& = (C \vee C) \wedge (C \vee A) \wedge (C \vee B) && \{\vee \text{ distributes over } \wedge\} \\
& = C \wedge (C \vee A) \wedge (C \vee B) && \{\vee \text{ idempotent}\} \\
& = C \wedge ((C \vee A) \wedge (C \vee B)) && \{\wedge \text{ associative}\} \\
& = C \wedge (C \vee (A \wedge B)) && \{\vee \text{ distributes over } \wedge\}
\end{aligned}
$$

33.

$$
\begin{aligned}
& C \vee A \wedge (B \vee C) \\
& = C \vee ((A \wedge B) \vee (A \wedge C)) && \{\wedge \text{ distributes over } \vee\} \\
& = C \vee ((A \wedge C) \vee (A \wedge B)) && \{\vee \text{ commutative}\} \\
& = (C \vee (A \wedge C)) \vee (A \wedge B) && \{\vee \text{ associative}\} \\
& = ((C \vee A) \wedge (C \vee C)) \vee (A \wedge B)) && \{\vee \text{ distributes over } \wedge\} \\
& = ((C \vee A) \wedge C) \vee A \wedge B && \{\vee \text{ idempotent}\}
\end{aligned}
$$

C.3   Predicate Logic

1.

$$(a) \quad F(1) \wedge F(2) \wedge F(3)$$
$$(b) \quad F(1) \vee F(2) \vee F(3)$$
$$(c) \qquad \big(G(1,1) \wedge G(1,2) \wedge G(1,3)\big)$$
$$\vee \big(G(2,1) \wedge G(2,2) \wedge G(2,3)\big)$$
$$\vee \big(G(3,1) \wedge G(3,2) \wedge G(3,3)\big)$$

2.

$$\big(F(1,5) \vee F(1,6)\big)$$
$$\wedge \big(F(2,5) \vee F(2,6)\big)$$
$$\wedge \big(F(3,5) \vee F(3,6)\big)$$
$$\wedge \big(F(4,5) \vee F(4,6)\big)$$

3.

- There is an even number.
  $\exists x. \ E(x)$

- Every number is either even or odd.
  $\forall x. \ (E(x) \vee O(x))$

- No number is both even and odd.
  $\forall x. \ \neg(E(x) \wedge O(x))$

- The sum of two odd numbers is even.
  $\forall x. \ \forall y. \ \big(O(x) \wedge O(y) \rightarrow E(x+y)\big)$

- The sum of an odd number and an even number is odd.
  $\forall x. \ \forall y. \ \big(E(x) \wedge O(y) \rightarrow O(x+y)\big)$

4.

- Chickens are birds.
  $\forall x. \ C(x) \rightarrow B(x)$

- Some doves can fly.
  $\exists x. \ D(x) \wedge F(x)$

- Pigs are not birds.
  $\forall x. \ P(x) \rightarrow \neg B(x)$

- Some birds can fly, and some can't.
  $\big(\exists x. \ B(x) \wedge F(x)\big) \wedge \big(\exists x. \ B(x) \wedge \neg F(x)\big)$

- An animal needs wings in order to fly.
  $\forall x.\ (\neg W(x) \rightarrow \neg F(x))$

- If a chicken can fly, then pigs have wings.
  $\big(\exists x.\ C(x) \wedge F(x)\big) \rightarrow \big(\forall x.\ P(x) \rightarrow W(x)\big)$

- Chickens have more feathers than pigs do.
  $\forall x.\ \forall y.\ (C(x) \wedge P(y)) \rightarrow M(x,y)$

- An animal with more feathers than any chicken can fly.
  $\forall x.\ \Big((A(x) \wedge (\forall y.\ (C(y) \wedge M(x,y)))) \rightarrow F(x)\Big)$

5.

- $\forall x.\ (\exists y.\ \text{wantsToDanceWith}\ (x,y))$
  Everybody has someone they want to dance with.

- $\exists x.\ (\forall y.\ \text{wantsToPhone}\ (y,x))$
  There is someone whom everybody wants to call.

- $\exists x.\ (\text{tired}\ (x) \wedge \forall y.\ \text{helpsMoveHouse}\ (x,y))$
  There is a person who is tired, and who helps everyone to move house.

10.

Theorem 74. $\forall x.F(x), \forall x.F(x) \rightarrow G(x) \vdash \forall x.G(x)$

Proof.

$$
\cfrac{
  \cfrac{\forall x.F(x)}{F(p)}\{\forall E\} \qquad \cfrac{\forall x.F(x) \rightarrow G(x)}{F(p) \rightarrow G(p)}\{\forall E\}
}{
  \cfrac{G(p)}{\forall x.G(x)}\{\forall I\}
}\{\rightarrow E\}
$$

$\square$

12.

Theorem 75. $\forall x.(F(x) \wedge G(x)) \vdash (\forall x.F(x)) \wedge (\forall x.G(x))$

Proof.

$$
\cfrac{
  \cfrac{\cfrac{\forall x.F(x) \wedge G(x)}{F(p) \wedge G(p)}\{\forall E\}}{\cfrac{F(p)}{\forall x.F(x)}\{\forall I\}}\{\wedge E_L\}
  \qquad
  \cfrac{\cfrac{\forall x.F(x) \wedge G(x)}{F(q) \wedge G(q)}\{\forall E\}}{\cfrac{G(q)}{\forall x.G(x)}\{\forall I\}}\{\wedge E_R\}
}{
  \forall x.F(x) \wedge \forall x.G(x)
}\{\wedge I\}
$$

$\square$

13. There will be 1000 terms containing F. In general, if the quantifiers are nested k deep, and the universe contains n elements, then the innermost term will occur n^k times.

C.4 Set Theory

13.

$(A' \cup B)' \cap C'$
$\quad = A'' \cap B' \cap C'$
$\quad = A \cap B' \cap C'$
$\quad = A \cap (B \cup C)'$

$A - (B \cup C)'$
$\quad = A \cap (B \cup C)''$
$\quad = A \cap (B \cup C)$

$(A \cap B) \cup (A \cap B')$
$\quad = A \cap (B \cup B')$
$\quad = A \cap U$
$\quad = A$

$A \cup (B - A)$
$\quad = A \cup (B \cap A')$
$\quad = (A \cup B) \cap (A \cup A')$
$\quad = (A \cup B) \cap U$
$\quad = A \cup B$

$A - B$
$\quad = A \cap B'$
$\quad = B' \cap A$
$\quad = B' \cap A''$
$\quad = B' - A'$

$(A \cap B) - (A \cap C)$
$\quad = (A \cap B) \cap (A \cap C)'$
$\quad = (A \cap B) \cap (A' \cup C')$
$\quad = ((A \cap B) \cap A') \cup ((A \cap B) \cap C')$
$\quad = (A \cap B \cap A') \cup (A \cap B \cap C')$
$\quad = \emptyset \cup (A \cap B \cap C')$
$\quad = A \cap B \cap C'$
$\quad = A \cap (B - C)$

$A - (B \cup C)$
$\quad = A \cap (B \cup C)'$
$\quad = A \cap (B' \cap C')$

$$= A \cap A \cap B' \cap C'$$
$$= (A \cap B') \cap (A \cap C')$$
$$= (A - B) \cap (A - C)$$

$A \cap (A' \cup B)$
$$= (A \cap A') \cup (A \cap B)$$
$$= \emptyset \cup (A \cap B)$$
$$= A \cap B$$

$(A - B') \cup (A - C')$
$$= (A \cap B'') \cup (A \cap C'')$$
$$= (A \cap B) \cup (A \cap C)$$
$$= A \cap (B \cup C)$$

14. Consider the expression $A \cap (B \cup C)$. It can be transformed as follows: $A \cap (B \cup C) = (A \cap B) \cup (A \cap C) = A \cap (B \cup C)$. The function would have to note any previous steps taken in order to avoid loops like this.

15. It is in simplest form.

16. $A \cup (C \cap A') = (A \cup C) \cap (A \cup A') = (A \cup C) \cap U = A \cup C$

17.

```
smaller :: Ord a => a -> [a] -> Bool
smaller x [] = True
smaller x (y:xs) = x < y

powerSet :: (Ord a, Eq a) => [a] -> [[a]]
powerSet set = normalizeSet (foldr g [[]] set)
   where g x acc =
              [x:epset | epset <- acc,
                  not (elem x epset) && smaller x epset]
              ++ acc
```

18. $(A \cup B)' = A' \cap B'$
$$= (U - A) \cap (U - B)$$
$$= ((A \cup A') - A) \cap ((B \cup B') - B)$$
$$= ((A \cup A') \cap A') \cap ((B \cup B') \cap B')$$

19.

```
isSubset :: Eq a => [a] -> [a] -> Bool
isSubset set1 set2 = null [e | e <- set1, not (elem e set2)]
```

20. The arguments are in the wrong order.

21. The second definition of e shadows the first, so the result is the second set.

22.

```
union :: Eq a => [a] -> [a] -> [a]
union set1 set2
  = set1 ++ [e | e <- set2, not (elem e set1)]
```

23. Yes, when A equals B and B and C are disjoint.

24. Both unions contain C, so C must appear in the intersection, unless C is empty and A and B are disjoint. Here is an example:

```
A = [1,2,3]
B = [4,5,6]
C = []
```

C.5   Recursion

1.

```
copy :: [a] -> [a]
copy []     = []
copy (x:xs) = x : copy xs
```

2.

```
inverse :: [(a,b)] -> [(b,a)]
inverse [] = []
inverse ((a,b):xs) = (b,a) : inverse xs
```

3.

```
merge :: Ord a => [a] -> [a] -> [a]
merge [] bs = bs
merge as [] = as
merge (a:as) (b:bs) =
  if a < b
    then a : merge as (b:bs)
    else b : merge (a:as) bs
```

4.

```
(!!) :: Int -> [a] -> Maybe a
(!!) n [] = Nothing
(!!) 0 (x:xs) = Just x
(!!) n (x:xs) = (!!) (n-1) xs
```

5.

```
lookup :: Eq a => a -> [(a,b)] -> Maybe b
lookup a [] = Nothing
lookup a ((a',b):ps)
  = if a==a' then Just b else lookup a ps
```

6.

```
countElts :: Eq a => a -> [a] -> Int
countElts e []      = 0
countElts e (x:xs) =
  if e == x
    then 1 + countElts e xs
    else countElts e xs
```

7.

```
removeAll :: Eq a => a -> [a] -> [a]
removeAll e [] = []
removeAll e (x:xs) =
  if e == x
    then removeAll e xs
    else x : removeAll e xs
```

8.

```
f :: [a] -> [a]
f [] = []
f (x:xs) = g xs

g :: [a] -> [a]
g [] = []
g (x:xs) = x:f xs
```

9.

```
extract :: [Maybe a] -> [a]
extract []           = []
extract (Nothing:xs) = extract xs
extract (Just x:xs)  = x : extract xs
```

10.

```
loop :: String -> String -> Int -> Maybe Int
loop [] s2 n = Nothing
loop (x:xs) s2 n
```

```
      = if length s2 > length (x:xs)
          then Nothing
          else if take (length s2) (x:xs) == s2
                  then Just n
                  else loop xs s2 (n+1)

  f :: String -> String -> Maybe Int
  f str1 str2 = loop str1 str2 0
```

11.

```
  foldrWith ::
    (a -> b -> c -> c) -> c -> [a] -> [b] -> c
  foldrWith f z [] bs = z
  foldrWith f z as [] = z
  foldrWith f z (a:as) (b:bs) =
    f a b (foldrWith f z as bs)
```

12.

```
  mappend :: (a -> [b]) -> [a] -> [b]
  mappend f xs = foldr fun [] xs
    where fun x acc = f x ++ acc
```

13.

```
  removeDuplicates :: Eq a => [a] -> [a]
  removeDuplicates [] = []
  removeDuplicates (x:xs)
    = if elem x xs
         then removeDuplicates xs
         else x : removeDuplicates xs
```

14.

```
  member :: Eq a => a -> [a] -> Bool
  member a []      = False
  member a (x:xs) = a == x || member a xs
```

15.

```
  appendTree :: Tree a -> [a] -> [a]
  appendTree Tip bs = bs
  appendTree (Node a t1 t2) bs =
    appendTree t1 (a : appendTree t2 bs)
```

16.

```
concatTree :: Tree [a] -> [a]
concatTree Tip = []
concatTree (Node as t1 t2) =
  concatTree t1 ++ as ++ concatTree t2
```

17.

```
zipTree :: Tree a -> Tree b -> Maybe [(a,b)]
zipTree Tip t2              = Nothing
zipTree t1  Tip             = Nothing
zipTree (Node a t1 t2) (Node b t3 t4)
 = case (zipTree t1 t3) of
     Nothing -> Nothing
     Just lst1 ->
       case (zipTree t2 t4) of
         Nothing -> Nothing
         Just lst2 -> Just (lst1 ++ [(a,b)] ++ lst2)
```

18.

```
zipWithTree ::
  (a -> b -> c) -> Tree a -> Tree b -> [c]
zipWithTree f Tip t2 = []
zipWithTree f t1 Tip = []
zipWithTree f (Node a t1 t2) (Node b t3 t4)
  = zipWithTree f t1 t3
    ++ [f a b]
    ++ zipWithTree f t2 t4
```

19.

```
intersection :: Eq a => [a] -> [a] -> [a]
intersection [] set2       = []
intersection (x:set1) set2
  = if (elem x set2)
       then x : intersection set1 set2
       else intersection set1 set2
```

20.

```
isSubset :: Eq a => [a] -> [a] -> Bool
isSubset [] set2     = True
isSubset (x:xs) set2 = elem x set2 /\ isSubset xs set2
```

21.

```
isSorted :: Ord a => [a] -> Bool
isSorted []      = True
isSorted (x:[]) = True
isSorted (x:y:xs) = x < y && isSorted (y:xs)
```

C.6   Inductively Defined Sets

20. The base case appears at the end of the stream, so cannot be used to start the inductive process of calculating successive elements. So, the stream does not have a printable value.

21. It will never terminate and return the accumulator in which the stream of naturals is constructed.

22. The powerset $P(S)$ is defined as follows:

- $\{\} \in P$

- $x \in S \wedge T \in P \rightarrow T \cup \{x\} \in P$

- Nothing is in P unless it can be shown to be in P by a finite number of uses of the base and induction rules.

Notice that each use of the induction case causes more than one set to be added to P. This is still inductive (we can order these sets by their sizes and so enumerate them).

23. No. The problem is that the stream may present an infinite number of naturals before the one we are interested in can be reached. For example, in `[1,3..]   ++  [2,4..]` all the odd naturals appear before any even number.

24. Given a number n, the set R of roots of n is defined as follows:

- $n^1 \in R$

- $n^{1/m} \in R \rightarrow n^{1/m+1} \in R$

- Nothing is in R unless it can be shown to be in R by a finite number of uses of the base and induction rules.

25.

1. $n^0 \in P$ by the base case

2. By the induction rule, $n^0 \in P \rightarrow n^1 \in P$ and so by *Modus Ponens*, $n^1 \in P$.

3. By the induction rule, $n^1 \in P \to n^2 \in P$ and so by *Modus Ponens*, $n^2 \in P$.

4. By the induction rule, $n^2 \in P \to n^3 \in P$ and so by *Modus Ponens*, $n^3 \in P$.

26. If n is a positive multiple of 2 then yes, otherwise no.

27. The odd positive integers.

28. By rule (1), $0 \in S$
By rule (2) $0 \in S \to 2 \in S$ and so by modus ponens $2 \in S$.
By rule (2) $2 \in S \to 4 \in S$ and so by modus ponens $4 \in S$.

29.

1. "" $\in YYS$

2. "" $\in YYS \to$ "yy" $\in YYS$ and so by *Modus Ponens*, "yy" $\in YYS$.

3. "yy" $\in YYS \to$ "yyyy" $\in YYS$ and so by *Modus Ponens*, "yyyy" $\in YYS$.

30.

```
zs = "" : map ('z':) zs
```

31. The set SS of strings of spaces of length less than or equal to n is defined as follows:

1. "" $\in SS$

2. $ss \in SS \wedge length\,ss < n \to$ ' '$||ss \in SS$

3. Nothing is in SS unless it can be shown to be in SS by a finite number of uses of rules (1) and (2).

32.

```
ss :: Int -> [String]
ss 0 = []
ss n = take n (repeat ' ') : ss (n-1)
```

33. The set of sets of integers SSI each of which is missing a distinct natural number is defined inductively as follows:

1. $I - \{0\} \in SSI$

2. $I - \{n\} \to I - \{n+1\} \in SSI$

3. Nothing is in SSI unless it can be shown to be in SSI by a finite number of uses of rules (1) and (2).

34.

   1. $I - \{-1\} \in SSI-$

   2. $I - \{-1\} \rightarrow I - \{-2\} \in SSI-$

   3. $I - \{-2\} \rightarrow I - \{-3\} \in SSI-$

   4. By *Modus Ponens*, $I - \{-3\} \in SSI-$.

35.

   1. $-1 \in ONI$

   2. $-1 \in ONI \rightarrow -3 \in ONI$ so $-3 \in ONI$.

   3. $-3 \in ONI \rightarrow -5 \in ONI$ so $-5 \in ONI$.

   4. $-5 \in ONI \rightarrow -7 \in ONI$ so $-7 \in ONI$.

36.

```
decrement :: Int -> Int
decrement x = x - 1

ni = -1 : map decrement ni
```

37. No, because 0 will be paired first with every element in `[0..]`, which is infinitely long.

38. The set is $\{0, 1\}$.

C.7   Induction

2. The sum of the first n odd numbers can be written as $\sum_{i=1}^{n}(2i-1)$, and we want to prove that this is equal to n^2. It's possible to prove this by induction, but a more elegant method is to reuse the theorem we already have, as follows:
$\sum_{i=0}^{n}(2i-1) = 2\sum_{i=0}^{n}i - \sum_{i=0}^{n}1 = 2\frac{n(n+1)}{2} - n = n(n+1) - n = n^2$.

3.

Proof. Induction over xs. The base case is:

   $length\ ([] +\!\!+ ys)$
     $= length\ ys$ $(+\!\!+).1$
       $= 0 + length\ ys$ $0 + x = x$
       $= length\ [] + length\ ys$ $length.1$

Assume that $length\ (xs +\!\!+ ys) = length\ xs + length\ ys$. The inductive case is:

$$length~((x:xs)+\!\!+ys)$$
$$= length~(x:(xs+\!\!+ys)) \qquad\qquad (+\!\!+).2$$
$$= 1 + length~(xs+\!\!+ys) \qquad\qquad length.2$$
$$= 1 + (length~xs~+~length~ys) \qquad\qquad \text{hypothesis}$$
$$= (1 + length~xs) + length~ys \qquad\qquad (+)~\text{is associative}$$
$$= length~(x:xs) + length~ys \qquad\qquad length.2$$

□

4.

Proof. Induction over xs. The base case is:

$$map~f~([]~+\!\!+~ys)$$
$$= map~f~ys \qquad\qquad (+\!\!+).1$$
$$= []~+\!\!+~map~f~ys \qquad\qquad (+\!\!+).1$$
$$= map~f~[]~+\!\!+~map~f~ys \qquad\qquad map.1$$

For the inductive case, assume $map~f~(xs+\!\!+ys) = map~f~xs~+\!\!+~map~f~ys$. Then

$$map~f~((x:xs)~+\!\!+~ys)$$
$$= map~f~(x:(xs+\!\!+ys)) \qquad\qquad (+\!\!+).2$$
$$= f~x~:~map~f~(xs+\!\!+ys) \qquad\qquad map.2$$
$$= f~x~:~(map~f~xs~+\!\!+~map~f~ys) \qquad\qquad \text{hypothesis}$$
$$= (f~x~:~map~f~xs)~+\!\!+~map~f~ys \qquad\qquad (+\!\!+).2$$
$$= map~f~(x:xs)~+\!\!+~map~f~ys \qquad\qquad map.2$$

□

5.

Proof. Using the definition of composition, we can rewrite the equation to be proved as:

$$map~f~(map~g~xs)~=~map~(f.g)~xs$$

We prove this by induction over xs. The base case is:

$$map~f~(map~g~[])$$
$$= map~f~[] \qquad\qquad map.1$$
$$= [] \qquad\qquad map.1$$
$$= map~(f.g)~[] \qquad\qquad map.1$$

Assume that $map~f~(map~g~xs)~=~map~(f.g)~xs$. The inductive case is:

$$map\ f\ (map\ g\ (x : xs))$$
$$\begin{aligned}
&= map\ f\ (g\ x\ :\ map\ g\ xs) &\quad map.2 \\
&= f\ (g\ x)\ :\ map\ f\ (map\ g\ xs) &\quad map.2 \\
&= f\ (g\ x)\ :\ map\ (f.g)\ xs &\quad \text{hypothesis} \\
&= ((f.g)\ x)\ :\ map\ (f.g)\ xs &\quad (.) \\
&= map\ (f.g)\ (x : xs) &\quad map.2
\end{aligned}$$

$\square$

7.

Theorem 76 (++ is associative). $(xs \mathbin{+\!\!+} ys) \mathbin{+\!\!+} zs = xs \mathbin{+\!\!+} (ys \mathbin{+\!\!+} zs)$

Proof. Induction over xs. The base case is:

$$([\,] \mathbin{+\!\!+} ys) \mathbin{+\!\!+} zs$$
$$\begin{aligned}
&= ys \mathbin{+\!\!+} zs \\
&= [\,] \mathbin{+\!\!+} (ys \mathbin{+\!\!+} zs)
\end{aligned}$$

Inductive case

$$((x : xs) \mathbin{+\!\!+} ys) \mathbin{+\!\!+} zs$$
$$\begin{aligned}
&= (x : (xs \mathbin{+\!\!+} ys)) \mathbin{+\!\!+} zs \\
&= x : ((xs \mathbin{+\!\!+} ys) \mathbin{+\!\!+} zs) \\
&= x : (xs \mathbin{+\!\!+} (ys \mathbin{+\!\!+} zs)) \\
&= (x : xs) \mathbin{+\!\!+} (ys \mathbin{+\!\!+} zs)
\end{aligned}$$

$\square$

8. The first line of the inductive case says 'Assume $P(n)$, and consider a set containing $n + 1$ horses; call them $h_1, h_2, \ldots, h_{n+1}$.' Consider what happens when $n = 1$. Our set of $n + 1$ horses contains only two of them, h_1 and h_2. Thus the two subsets turn out to be $A = \{h_1\}$ and $B = \{h_2\}$. All the horses in A have the same colour, and all the horses in B have the same colour. So far, so good—but the next sentence says 'Pick one of the horses that is an element of both A and B,' *and it is impossible to do this because when $n = 1$ the sets A and B are disjoint.* The rest of the proof is invalid because it relies on this non-existent horse.

There are two useful lessons to learn from this.

- Whenever a proof says to 'pick an x such that $\ldots$', it is essential to make sure that such an x exists.

- It is helpful to work through the details using a concrete example.

By the way: the flaw explained above is the only error in this proof. *If* the proof worked for the case $n = 1$—if all pairs of horses had the same colour—then *it would indeed be true* that all horses are the same colour.

9.

$reverse\ (reverse\ [1, 2, 3])$
$= reverse\ [3, 2, 1]$
$= [1, 2, 3]$

10.

Proof. Induction over xs. The base case:

$reverse([]\mathbin{+\!\!+} ys)$
$= reverse\ ys$
$= reverse\ ys\mathbin{+\!\!+}[]$
$= reverse\ ys\mathbin{+\!\!+} reverse\ []$

Inductive case:

$reverse\ ((x : xs)\mathbin{+\!\!+} ys)$
$= reverse\ (x : (xs\mathbin{+\!\!+} ys))$
$= reverse\ (xs\mathbin{+\!\!+} ys)\mathbin{+\!\!+}[x]$
$= (reverse\ ys\mathbin{+\!\!+} reverse\ xs)\mathbin{+\!\!+}[x]$
$= reverse\ ys\mathbin{+\!\!+}(reverse\ xs\mathbin{+\!\!+}[x])$
$= reverse\ ys\mathbin{+\!\!+}(reverse(x : xs))$

$\square$

11.

Proof. Induction over xs. The base case is:

$reverse\ (reverse\ [])$
$= reverse\ []$
$= []$

For the inductive case, assume for some xs that $reverse\ (reverse\ xs) = xs$. Then

$reverse\ (reverse\ (x : xs))$
$= reverse\ (reverse\ xs\mathbin{+\!\!+}[x])$
$= reverse\ [x]\mathbin{+\!\!+} reverse\ (reverse\ xs)$
$= [x]\mathbin{+\!\!+} xs$
$= x : xs$

$\square$

12. Consider, for example, the infinite list *nats* $=$ $[0, 1, 2, 3, \ldots]$. This is easily defined in Haskell, with the expression [0..]. Notice that *nats* is not an infinite loop; a computation requiring only a finite portion of it will terminate. On the other hand, any computation requiring all of *nats* will go into an infinite loop; thus *length nats* will never terminate, and we can express this by saying *length nats* $=$ $\bot$.

Now consider the equation *reverse (reverse nats)* $=$ *nats*. The left hand side of the equation is not just a vague, intuitive statement that might be interpreted as leaving a list unchanged. It has a specific meaning which can (and must) be determined by equational reasoning using the definition of *reverse*.

The outermost application of reverse must begin by determining which of the two defining equations is relevant; it does this by performing a case analysis on its argument (which is *reverse nats*) to decide whether this is the empty list [] or a non-empty list in the form x:xs.

$reverse \ (reverse \ (nats \ 0))$
  $= \textbf{case} \ reverse \ (nats \ 0) \ \textbf{of}$
    $[] \rightarrow []$
    $x : xs \rightarrow reverse \ xs \mathbin{+\!\!+} [x]$

Now the computer must evaluate *reverse (nats 0)* far enough to decide whether it is the empty list, or a cons expression. This evaluation proceeds as follows:

$reverse \ (nats \ 0)$
  $= reverse \ (0 : nats \ 1)$
  $= reverse \ (nats \ 1) \mathbin{+\!\!+} [0]$
  $= reverse \ (1 : nats \ 2) \mathbin{+\!\!+} [0]$
  $= (reverse \ (nats \ 2) \mathbin{+\!\!+} [1]) \mathbin{+\!\!+} [0]$

This evaluation is never going to terminate. The computer will just keep generating larger and larger natural numbers, constructing ever bigger expressions, but it will never actually figure out whether the original value *reverse (nats 0)* is empty! Hence no information at all can be obtained from evaluating

$$reverse \ (reverse \ (nats \ 0)).$$

To summarise, we know that:

$$reverse \ (reverse \ (nats \ 0)) \quad = \quad \bot$$
$$nats \ 0 \quad \neq \quad \bot$$

A consequence of this, which might be surprising, is that

$$reverse \ . \ reverse \ \neq \ id.$$

C.8 Relations

45.

- (a) Yes.

- (b) This depends upon whether the relationship is defined by DNA, in which case all of us are somewhat related to each other, or last name, in which case some people named Smith are not actually closely related physically. 'No' seems a reasonable answer.

- (c) No, because equivalence relations are reflexive, and you cannot be bigger than yourself.

- (d) Yes.

46. Yes. Here is an example:

```
[(1,2),(2,3),(3,1)]
```

47. The empty relation.

48.

```
checkPowers :: (Eq a, Show a) => Digraph a -> Bool
checkPowers (set,relation)
 = any (setEq relation)
       [relationalPower (set,relation) n
         | n <- [2..length (domain relation)]]
```

49. No. If it did, then the end of each loop would have to have a reflexive loop, since the relation is transitive. Thus it would not be irreflexive.

50. Yes. For example

```
[(1,2),(2,3),(3,1)]
```

has a transitive closure that is reflexive and symmetric.

51.

```
fewerArcs :: (Eq a, Show a) => Digraph a -> Bool
fewerArcs (set,relation)
  = all
     (< (length relation))
        [length (relationalPower (set,relation) n)
           | n <- [2..length (domain relation)]]
```

52.

```
isSmaller :: (Ord a, Show a) => Digraph a -> Bool
isSmaller (set,relation)
  = let (symset,symrelation) =
          symmetricClosure (set,relation)
    in length relation < length symrelation
```

53. The last is not a topological sort.

54. Yes, for example this is reflexive, symmetric and antisymmetric:

```
[(1,1),(2,2),(3,3)]
```

55.

```
numTransArcs :: Digraph a -> Int
numTransArcs (set,relation) =
  sum [1..length relation - 1]
```

56. The number of arcs the transitive closure will add is $1+2+\ldots+n-1$. The symmetric closure will double each of these, so the total is $2(1+2+\ldots+n-1)$.

57. Power n.

58. No, because the function could not examine every arc in the relation.

59. Given a partial order, assume that it has a cycle of length n. Since it is transitive, it also has cycles of lengths 1 to $n-1$. But that means that it has a cycle of length 2, which cannot be because a partial order is anti-symmetric. So it cannot both be a partial order and have cycles greater than 1.

60. No. Some have powers that repeat, such as the relation

```
[(1,2),(2,1)]
```

61. A linear order.

62. Yes. For example, the composition of these two relations

```
[(Red,Blue),(Blue,Green),(Green,Yellow),(Yellow,Red)]
```

```
[(Blue,Red),(Green,Blue),(Yellow,Green),(Red,Yellow)]
```

yields

```
[(Red,Red),(Blue,Blue),(Green,Green),(Yellow,Yellow)]
```

63. The empty relation and the equality relation.

C.9   Functions

26. The functions surjExp, injExp, and bijExp each take two domains and images and test the corresponding hypothesis.

```
surjectiveExperiment :: (Eq a, Show a) =>
 Set a -> Set a -> Set (a, FunVals a) ->
 Set a -> Set a -> Set (a, FunVals a) -> Bool
surjectiveExperiment dom_f co_f f dom_g co_g g
 = isSurjective dom_f co_f f /\
   isSurjective dom_g co_g g
   ==>
   isSurjective dom_g co_f
   (functionalComposition f g)

surjExp :: Set Int -> Set Int ->
           Set Int -> Set Int -> Bool
surjExp domain_f image_f domain_g image_g
 = let f = [(x,Value y) | x <- domain_f, y <- image_f]
       g = [(x,Value y) | x <- domain_g, y <- image_g]
   in
   surjectiveExperiment [1..10] [1..10] f
                        [1..10] [1..10] g

injectiveExperiment :: (Eq a, Show a) =>
 Set a -> Set a -> Set (a, FunVals a) ->
 Set a -> Set a -> Set (a, FunVals a) -> Bool
injectiveExperiment dom_f co_f f dom_g co_g g
 = isInjective dom_f co_f f /\
   isInjective dom_g co_g g
   ==>
   isInjective dom_g co_f
   (functionalComposition f g)

injExp :: Set Int -> Set Int ->
          Set Int -> Set Int -> Bool
injExp domain_f image_f domain_g image_g
 = let f = [(x,Value y) | x <- domain_f, y <- image_f]
       g = [(x,Value y) | x <- domain_g, y <- image_g]
   in injectiveExperiment [1..10] [1..10] f
                          [1..10] [1..10] g

bijectiveExperiment :: (Eq a, Show a) =>
 Set a -> Set a -> Set (a, FunVals a) ->
 Set a -> Set a -> Set (a, FunVals a) -> Bool
bijectiveExperiment dom_f co_f f dom_g co_g g
```

```
        = isBijective dom_g co_f
          (functionalComposition f g)
          ==>
          isSurjective dom_f co_f f /\
          isInjective dom_g co_g g

    bijExp :: Set Int -> Set Int ->
              Set Int -> Set Int -> Bool
    bijExp domain_f image_f domain_g image_g
        = let f = [(x,Value y) | x <- domain_f, y <- image_f]
              g = [(x,Value y) | x <- domain_g, y <- image_g]
          in bijectiveExperiment [1..10] [1..10] f
                                 [1..10] [1..10] g
```

27. 1,1,4,4

28. The function f is a permutation, g is an identity function and h is the inverse of f.

29. constant function, injection, constant function.

30. Only g is a function. f maps 3 to two values, and g does not map 5 to any value.

31. g is a total function, and the other two are partial functions.

32.

```
    h o h      yes
    f o g      no
    g o f      yes
    h o f      domains don't match
    g o h      domains don't match
```

33.

```
    f o g      no
    g o f      yes
    h o f      no
```

34.

```
    g o g      yes
    h o f      no
    f o h      yes
```

35. The first function is partial, because it does not terminate when given
`False`, and so does not produce a result given that domain value. The second
function is total.

36.

```
isFun :: (Eq a, Show a) => Set (a,b) -> Bool
isFun ps = normalForm (map fst ps)
```

37.

```
isInjection :: (Eq a, Eq b, Show a, Show b)
        => Set (a,b) -> Bool
isInjection ps
  = normalForm (map fst ps) /\ normalForm (map snd ps)
```

39. You only need to know its type. If the domain type is the same as that
of the codomain, then it might be the identity function, otherwise it certainly
isn't. There are many properties of functions that can be deduced solely from
the function type.

40.

```
compare f g h = ((g.f) a) == (h a)
```

41. Yes, for natural numbers. If it receives a negative integer, it will loop
indefinitely.

42. Yes, for all integers.

C.10 Discrete Mathematics in Circuit Design

1.

$$
\begin{aligned}
f\ a\ b\ c =\ & \\
& (\neg a \wedge \neg b \wedge \neg c) \\
\vee\ & (\neg a \wedge \neg b \wedge c) \\
\vee\ & (a \wedge \neg b \wedge c) \\
\vee\ & (a \wedge b \wedge \neg c) \\
\vee\ & (a \wedge b \wedge c)
\end{aligned}
$$

8.

```
Example: addition of 13+41=54 using 6-bit words
   13 = 001101
   41 = 101001
  sum = 110110 = 54, with carry out = 0
   rippleAdd False [(False,True),(False,False),(True,True),
                (True,False),(False,False),(True,True)]
The expected result is
   (False, [True,True,False,True,True,False])
```

9.

```
Test cases for halfCmp
halfCmp (False,False) -- (False,True,False)  since x=y
halfCmp (False,True) -- (True,False,False)  since x<y
halfCmp (True,False) -- (False,False,True)  since x>y
halfCmp (True,True) -- (False,True,False)  since x=y
```

10.

```
Test cases for rippleCmp
rippleCmp [(False,False),(False,True),(True,False)]
    -- (True,False,False) since x<y
rippleCmp [(False,False),(True,True),(False,False)]
    -- (False,True,False) since x=y
rippleCmp [(False,False),(True,False),(False,True)]
    -- (False,False,True) since x>y
```

12.

```
Test cases for the full adder, with expected results
   fullAdd (False,False) False   -- 0 0
   fullAdd (False,True) False   -- 0 1
   fullAdd (True,False) False   -- 0 1
   fullAdd (True,True) False   -- 1 0
   fullAdd (False,False) True   -- 0 1
   fullAdd (False,True) True   -- 1 0
   fullAdd (True,False) True   -- 1 0
   fullAdd (True,True) True   -- 1 1
```

17. You can't test an adder thoroughly because there is an exponential growth in the size of the truth table as the word size grows. For an n-bit word, there are 2×2^{2n} lines in the truth table. The exponent is $2n$ because there are two bits for each position, and the initial factor of 2 accounts for the possibility that the carry input is either 0 or 1. For a current generation processor, where the word size is 64, the truth table would have 2^{129} lines, which is larger than the size of the known universe (it hardly matters what unit of measurement you choose!)

Bibliography

[1] Harold Abelson, Gerald Jay Sussman, and Julie Sussman. *Structure and Interpretation of Computer Programs*. The MIT Press, 1985.

[2] Martin Aigner and Günter M. Ziegler. *Proofs from THE BOOK*. Springer, 1998.

[3] Richard Bird. *Introduction to Functional Programming using Haskell*. Prentice Hall, second edition, 1998.

[4] Richard Bird and Oege de Moor. *Algebra of Programming*. Prentice Hall, 1997.

[5] Stanley N. Burris. *Logic for Mathematics and Computer Science*. Prentice Hall, 1998.

[6] C. J. Date. *An Introduction to Database Systems*. Addison Wesley Longman, 1999.

[7] Antony Davie. *An Introduction to Functional Programming Systems Using Haskell*. Cambridge University Press, 1992.

[8] Michael Downward. *Logic and Declarative Language*. Taylor & Francis, 1998.

[9] Herbert B. Enderton. *A Mathematical Introduction to Logic*. Academic Press, 1972.

[10] Herbert B. Enderton. *Elements of Set Theory*. Academic Press, 1977.

[11] Arthur Engel. *Problem-Solving Strategies*. Springer, 1998.

[12] Jean-Yves Girard, Yves Lafont, and Paul Taylor. *Proofs and Types*, volume 7 of *Cambridge Tracts in Theoretical Computer Science*. Cambridge University Press, 1989.

[13] Derek Goldrei. *Classic Set Theory*. Chapman & Hall, 1996.

[14] Ronald L. Graham, Donald E. Knuth, and Oren Patashnik. *Concrete Mathematics*. Addison-Wesley, 1989.

[15] Jan Gullberg. *Mathematics from the Birth of Numbers*. Norton, 1997.

[16] Douglas R. Hofstadter. *Gödel, Escher, Bach: An Eternal Golden Braid*. Penguin, 1979.

[17] Paul Hudak. *The Haskell School of Expression: Learning Functional Programming through Multimedia*. Cambridge University Press, 2000.

[18] Geraint Jones and Mary Sheeran. Relations and refinement in circuit design. In *3rd Refinement Workshop*, Workshops in Computing. Springer, 1991.

[19] Donald E. Knuth, Tracy Larrabee, and Paul M. Roberts. *Mathematical Writing*. Number 14 in MAA Notes. The Mathematical Association of America, 1989.

[20] E. J. Lemmon. *Beginning Logic*. Chapman & Hall, 1965.

[21] John O'Donnell. From transistors to computer architecture: Teaching functional circuit specification in hydra. In *FPLE'95: Symposium on Functional Programming Languages in Education*, volume 1022 of *LNCS*, pages 195–214. Springer, December 1995.

[22] Chris Okasaki. *Purely Functional Data Structures*. Cambridge University Press, 1998.

[23] John O'Leary, Xudong Zhao, Rob Gerth, and Carl-Johan Seger. Formally verifying IEEE compliance of floating-point hardware. *Intel Technology Journal*, 1999.

[24] Kenneth A. Ross and Chalres R. B. Wright. *Discrete Mathematics*. Prentice Hall, 1999.

[25] Simon Singh. *The Code Book: The Science of Secrecy from Ancient Egypt to Quantum Cryptography*. Fourth Estate, 1999.

[26] Raymond Smullyan. *Forever Undecided: A Puzzle Guide to Gödel*. Oxford University Press, 1987.

[27] Donald F. Stanat and David F. McAllister. *Discrete Mathematics in Computer Science*. Prentice Hall, 1977.

[28] Patrick Suppes. *Axiomatic Set Theory*. Dover, 1972.

[29] Simon Thompson. *Type Theory and Functional Programming*. Addison-Wesley Publishing Company, 1991.

[30] Simon Thompson. *Haskell: The Craft of Functional Programming*. Addison-Wesley, second edition, 1999.

[31] Daniel J. Velleman. *How To Prove It: A Structured Approach*. Cambridge University Press, 1994.

Index